Cell and Developmental Biology of the Eye

Cell and Developmental Biology of the Eye

Series Editors
Joel B. Sheffield and S. Robert Hilfer

Ocular Size and Shape: Regulation During Development

Cellular Communication During Ocular Development

Molecular and Cellular Basis of Visual Acuity

Heredity and Visual Development

Development of Order in the Visual System

The Proceedings of the Philadelphia Symposia on Ocular and Visual Development

Development of Order in the Visual System

Edited by
S. Robert Hilfer and
Joel B. Sheffield

With 96 Figures

Springer-Verlag
New York Berlin Heidelberg Tokyo

S. Robert Hilfer
Joel B. Sheffield
Department of Biology
Temple University
Philadelphia, Pennsylvania 19122, U.S.A.

On the cover: SEM micrographs of cortical fibers. See page 43.

Library of Congress Cataloging-in-Publication Data
Main entry under title:
Development of order in the visual system.
(Cell and developmental biology of the eye)
Papers from the Ninth Symposium on Ocular and Visual Development, held in Oct. 1984, in Philadelphia.
Includes bibliograhies and index.
1. Visual pathways—Congresses. 2. Neural circuitry—Congresses. 3. Developmental neurology—Congresses. 4. Cell interaction—Congresses. 5. Eye—Congresses. I. Sheffield, Joel B. II. Hilfer, S. Robert. III. Symposium on Ocular and Visual Development (9th : 1984 : Philadelphia, Pa.) IV. Series. [DNLM: 1. Visual Perception—physiology—congresses. W3 SY5363 9th 1984d WW 105 D4886 1984] QP475.D48 1986 599'.01823 85-27820

Printed and bound by Halliday Lithograph Corporation,
West Hanover, Massachusetts
Printed in the United States of America.
9 8 7 6 5 4 3 2 1
ISBN 0-387-96264-6 Springer-Verlag New York Berlin Heidelberg Tokyo
ISBN 3-540-96264-6 Springer-Verlag Berlin Heidelberg New York Tokyo

Series Preface

The eye has fascinated scientists from the earliest days of biological investigation. The diversity of its parts and the precision of their interaction make it a favorite model system for a variety of developmental studies. The eye is a particularly valuable experimental system not only because its tissues provide examples of fundamental processes, but also because it is a prominent and easily accessible structure at very early embryonic ages.

In order to provide an open forum for investigators working on all aspects of ocular development, a series of symposia on ocular and visual development was initiated in 1973. A major objective of the symposia has been to foster communication between the basic research worker and the clinical community. It is our feeling that much can be learned on both sides from this interaction. The idea for an informal meeting allowing maximum exchange of ideas originated with Dr. Leon Candeub, who supplied the necessary driving force that made the series a reality. Each symposium has concentrated on a different aspect of ocular development. Speakers have been selected to approach related topics from different perspectives.

This book series, "Cell and Developmental Biology of the Eye," is derived from the Philadelphia symposia on ocular and visual development. Previous volumes are listed on the series page. We hope that the introduction of this proceedings series will make the results of research on ocular cell and developmental biology more widely known and more easily accessible.

Preface

The ninth Symposium on Ocular and Visual Development was held in October, 1984, to explore aspects of the role of order in visual development. Although there is enormous apparent plasticity in visual function, studies of anatomy, physiology, and behavior have revealed a striking amount of underlying order. This order is perhaps best exemplified by the structure of the vertebrate lens, or the compound eye of invertebrates, but is also apparent in the array of neural processes in visual centers of the brain. One can probe the functional consequences of underlying order with behavioral studies on stereopsis and pattern recognition which attempt to discern ordered restrictions on visual functions. Conversely, it is informative to examine the induction and potential reversal of amblyopia to investigate the contributions of neural activity to the establishment and maintenance of functional order.

We are indebted to our colleagues Dr. E. Gruberg from Temple University and Dr. L. Andrews from the Pennsylvania College of Optometry for their expertise and assistance in the organization of the meeting, to the speakers for their presentations and contributions to this volume, and to the reviewers of the manuscripts for their helpful comments.

This symposium could not have been held without the generous support of the Temple University College of Arts and Sciences and the Pennsylvania College of Optometry. We also thank Merck, Sharp and Dohme, Inc., Alcon Foundation, and The March of Dimes Birth Defects Corporation for their support. This volume was prepared with the skillful assistance of Huong Tran to whom we are indebted.

September 20, 1985
Philadelphia, Pennsylvania

S. Robert Hilfer
Joel B. Sheffield

Contents

Contributors

Chiye Aoki, The Rockefeller University, New York, NY 10021, U.S.A.

B.A. Bertram, Departments of Pathology, Physiology, and Ophthalmology, Rush Presbyterian St. Lukes' Medical Center, Chicago, IL 60612, U.S.A.

Kenneth J. Ciuffreda, Amblyopia Laboratory, Institute for Vision Research, SUNY/State College of Optometry, New York, NY 10010, U.S.A.

M. Constantine-Paton, Department of Biology, Yale University, P.O. Box 6666, New Haven, CT 06511, U.S.A.

Joseph M. Corless, Departments of Anatomy and Ophthalmology, Duke University Medical Center, Durham, NC 27710, U.S.A.

M.L.J. Crawford, University of Texas Graduate School of Biomedical Sciences, Houston, TX 77025, U.S.A.

R.W. Cummings, Pennsylvania College of Optometry, William Feinbloom Low Vision Rehabilitation Center, Philadelphia, PA 19140, U.S.A.

Paul G. FitzGerald, Department of Human Anatomy, School of Medicine, University of California, Davis, CA 95616, U.S.A.

Bela Julesz, AT&T Bell Laboratories, Murray Hill, NJ 07974, U.S.A.

J.R. Kuszak, Departments of Pathology, Physiology, and Ophthalmology, Rush Presbyterian St. Lukes' Medical Center, Chicago, IL 60612, U.S.A.

Richard M. Lebovitz, Department of Biology, Princeton University, Princeton, NJ 08540, U.S.A.

J.J. Norden, Department of Anatomy, Vanderbilt University, School of Medicine, Nashville, TN 37232, U.S.A.

J.L. Rae, Departments of Pathology, Physiology, and Ophthalmology, Rush Presbyterian St. Lukes' Medical Center, Chicago, IL 60612, U.S.A.

Donald F. Ready, Department of Biology, Princeton University, Princeton, NJ 08540, U.S.A.

Philip Siekevitz, The Rockefeller University, New York, NY 10021, U.S.A.

Andrew Tomlinson, Department of Biology, Princeton University, Princeton, NJ 08540, U.S.A.

Susan Boymel Udin, Division of Neurobiology, Department of Physiology, State University of New York, Buffalo, NY 14214, U.S.A.

S.G. Whittaker, Pennsylvania College of Optometry, William Feinbloom Low Vision Rehabilitation Center, Philadelphia, PA 19140, U.S.A.

Synapse Regulation in the Developing Visual System

M. Constantine-Paton and J.J. Norden

The number and distribution of synaptic contacts between two neuronal populations is an obviously important component of the strength and kind of interaction between the two sets of cells. Nevertheless, little is known about how characteristic innervation densities and distributions are controlled during development. At one extreme is the possibility that these parameters are set by the number and type of neurons in the target cell population. At the other extreme synaptic innervation density and distribution could be directly related to the number of afferent neurons projecting to a given target area.

We have been addressing this question using the visual pathway of the frog Rana pipiens. In this animal, it is possible to implant surgically a third eye primordium at an early neural tube stage of development. The embryonic graft differentiates into a morphologically normal eye that sends a supernumerary optic nerve to coinnervate an optic tectal lobe along with the retinal projection from one of the original eyes of the host. The striking characteristic of the tecta that become double-innervated as a result of the procedure is that the normal continuous visual projection (from the contralateral host retina) becomes periodically disrupted into a pattern of eye specific stripes that alternate with stripes representing the recipient zones of synaptic terminals from the supernumerary retina (figure 1). Despite this disruption of continuity, however, the embryonically appropriate regions of both retinas continue to terminate in roughly appropriate regions of the double-innervated tectal lobe (Constantine-Paton and Law, 1978). The induced pattern is in fact remarkably similar to the patterns of ocular dominance columns found in the visual cortex and superior

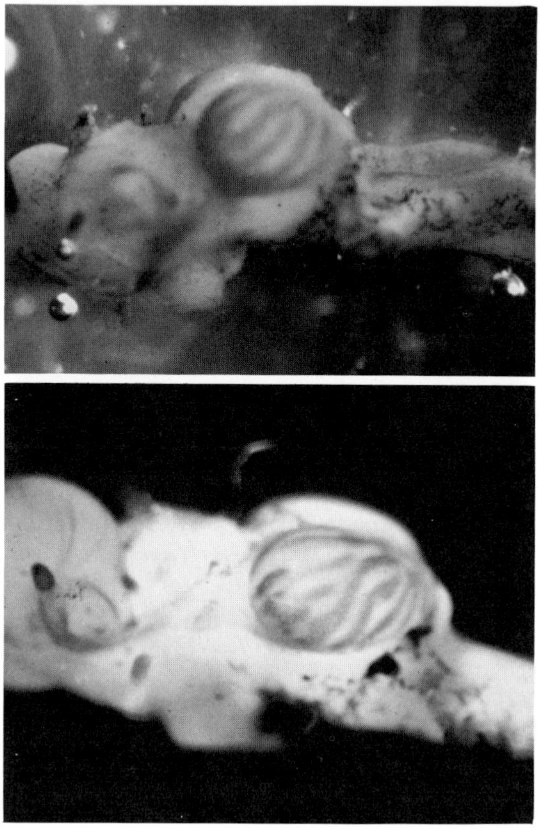

Figure 1. Rostro-caudal orientation of eye-specific stripes. Brains of 2 three-eyed tadpoles (T & K Stage VI) processed for visualization of the enzyme horseradish peroxidase (HRP) reaction product after the enzyme was used to label the tectal projection sites of one of the two doubly innervating optic tracts. The photographs show the stereotyped rostral to caudal orientation of the striped termination zones and the similar periodicity of the stripes in the right tectal lobes of the two brains even though details of the pattern vary between the two animals.

colliculus of monkeys where input from two eyes normally converges on the same target cell population (LeVay et al., 1975; Pollack and Hickey, 1979).

A "COMPROMISE" HYPOTHESIS FOR MAPS AND STRIPES

We have suggested that this striped pattern reflects two completely independent mechanisms that normally serve to bring about normal topographic maps not only in the frog but in all of the vertebrate central nervous system (CNS) pathways in which they are known to occur. The first involves a form of biochemical matching between retinal ganglion cells (RGCs) and tectal neurons by virtue of their relative positions in the retinal or tectal cell array. This positional matching serves to distribute the retinal axon populations with a fixed alignment throughout the retinal terminal zone of the tectum. A second "sorting" mechanism probably only operates once functional synapses are established. It could function normally to sharpen the precision of the retinal projection through a process that stabilizes synaptic contacts only if they are located adjacent to other synapses arising from retinal neighbors. When two separate retinas (or any two presynaptic populations) are forced to coinnervate one target zone these mechanisms working in concert are capable of producing interdigitated elongated terminal zones. The requirement for positional matching will distribute both presynaptic axon populations throughout the target. The sorting mechanism will produce periodic source specific stripes because the pattern represents an optimal compromise. The developing synapses are able to maximize the number of contacts they make in the appropriate part of the target while also minimizing the number of times they have to make contacts adjacent to other synapses that arise from cells that are not their presynaptic neighbors (Law and Constantine-Paton, 1981a).

The explanation given above has been supported by a number of experiments. Through a variety of embyonic manipulations, it has been shown that only the embryonic retinal position of doubly innervating ganglion cells and the neighborhood relationships of doubly innervating ganglion cell bodies are critical to formation of the striped termination pattern (Law and Constantine-Paton, 1980; 1981a: Fawcett and Willshaw, 1982; Constantine-Paton et al., 1983; Ide et al., 1983). Moreover, we now know that the mechanism responsible for synapse sorting on the basis of ganglion cell body proximity is dependent on a normal pattern of action potential activity in the ganglion cell population (Meyer, 1982, 1983; Schmidt and Edwards, 1983; Reh and Constantine-Paton, 1984). This supports the idea that high temporal correlations in the pattern of action potentials among neighboring ganglion cells provide the critical proximity information. (Arnett, 1978; Arnett and Spraker, 1981; Mastronarde, 1983). A number of diverse experiments have suggested that post-synaptic neurons may have the ability to selectively stabilize those inputs that effectively drive them through temporal summation of converging activity (Hubel and Wiesel, 1965; Stent, 1973; Changeux and Danchin, 1976). This interaction could produce the second refinement stage of map formation that produces incidentally segregation of inputs from different presynaptic sources. Thus in a retinal projection convergence of terminals from two synchronously active retinal neighbors would be favored over inputs from distantly positioned cells because activity in the retinal neighbors could summate to effectively drive the tectal neuron and produce stabilization of these synapses. Activity from non-neighbors would seldom summate and consequently these inputs would ultimately be replaced.

One troublesome aspect of the preceding explanation is that neither the position matching component nor the activity-dependent component can readily account for two major properties of the segregation patterns that occur normally in mammals or that are induced upon double innervation of a tectum in frogs. Namely, eye-specific stripes always run in a rostrolateral to caudomedial direction and the periodicity of the stripes is very regular. In the frog, the period is always between 300 and 400 microns in animals of very different ages and very different treatments.

QUANTITATIVE ANALYSES OF DOUBLE INNERVATION

As a means of exploring the numerical determinants of the stereotypy in the experimentally produced termination pattern, we began quantitative light microscope studies of retinal ganglion cell numbers, tectal volume changes and neuropil volumes occupied by terminals from each retina. We also asked whether the number or type of synaptic contacts is altered in the doubly versus the singly innervated optic tectal lobe of the same animal. Our results led us to some unexpected conclusions regarding the respective roles played by tectal cells or circuits and retinal afferents in patterning the retinotectal connection.

A summary of the analysis of RGC number is presented in figure

2. The graphs demonstrate that double innervation of an optic tectum in the frog tadpole does not result in increased retinal ganglion cell death. This is in marked distinction to the expected result in any warm-blooded vertebrate where a number of investigators have shown that the degree of competition for terminal space in central visual targets can be related to the number of retinal ganglion cells that die during maturation (see Cowan et al., 1984 for a recent review). Lack of similar lability in the frog may be a reflection of an inherently higher tolerance in cold-blooded vertebrate retinal ganglion cells for lack of trophic support from a target. Even in normal tadpoles there is no sign of the ontogenetic retinal ganglion cell death seen in chicks and mammals (Cunningham, 1982; Reh and Constantine-Paton, 1983) and it is well established that fish and frog retinal ganglion cell axons have a far greater regenerative ability when injured.

A number of studies conducted over the past 10 years have indicated that retinal ganglion cells from embryonically manipulated eyes seldom make significant numbers of synaptic contacts in abnormal central targets. Moveover, if the optic tectum is made completely inaccessible to the growing optic tract even frog retinal ganglion cells will die in large number (Constantine-Paton and Capranica, 1976; Reh et al., 1983). Consequently, it can be assumed that the vast majority of ganglion cells counted in the retinas of three-eyed frogs are making contacts successfully with the tectal lobe. Thus, the finding that retinal ganglion cell number is not consistently reduced in any of the eyes of these animals implies that, in general, striped tectal lobes receive input from twice the normal number of retinal neurons (Law and Constantine-Paton, 1981b).

Volume measurements performed on the normal and striped tectal lobes of the same animals suggest that the tectum itself constitutes a critical limiting factor in the numerical adjustments forced by implanting the supernumerary eye. Comparisons of total tectal volumes revealed an average hypertrophy of 26% in the doubly innervated lobe. Even measurements restricted to the superficial retinal recipient zone of the tectal lobes failed to reveal increases in this central target that were equal to the striped tectum, and the supernumerary eye's projection to the striped tectum indicated that the coinnervating projections were effectively compressed into 69% of their normal neuropil space (Constantine-

Paton and Ferrari-Eastman, 1985).[1]

Stereological electron microscopic analyses conducted in parallel to the quantitative light microscopic studies also support the idea that the optic tectum plays the determinative role in controlling the numerical balance of pre- to post-synaptic inputs. The initial analyses in this investigation were performed without the microscopist knowing which tectal lobe received the double retinal innervation. Identification of the doubly innervated tectum was accomplished in adjacent 1 or 2 micron sections processed for light microscope autoradiography. The anterograde transport of ^3H proline from an eyecup was employed to label the striped innervation zone from the supernumerary eye in two post-metamorphic (1 mo. post-metamorphosis) frogs. These analyses (see table 1) reveal that when corrected for the hypertrophy which occurs in the doubly innervated

Figure 2. Comparisons of retinal ganglion cell counts in normal and dually innervating eyes. (A) Comparisons of normal and supernumerary eyes: Total retinal ganglion cell (RGC) counts from the transplanted eyes of 9 three-eyed animals (●) are plotted along with equivalent data from total counts of retinas in 22 normal binocular frogs (•). Animals ranged in age from young limb-bud stage tadpoles (T & K Stage IV) to young post-metamorphic frogs (T & K Stage XXV; one week past emergence onto land). In both sets of animals total RGC counts increase with age as the eye grows. Ganglion cell counts in the normal animals. (B) Comparisons of RGC counts in the host animals' retinas. Total RGC counts are plotted against age for the 8 three-eyed animals in which cell counts were obtained for both host retinas. It can be seen that the host retina projecting to the doubly innervated tectum (▲) is not consistently reduced in total RGC number compared to the host eye which projected to the predominantly "normal" tectum of the same animal (△). (C) Comparisons of the number of RGCs in the supernumerary eyes (●) and in the competing host retinas (▲). It can be seen that there is no consistent reduction in the number of RGCs in the transplanted eye compared to the host eye with which it must compete for synaptic space in the doubly innervated tectum. (D) Comparison of RGCs in the supernumerary eyes (●) and in the predominantly non-competing (△) host eyes. Again there is no consistent telationship between the cell counts and the amount of centroal synaptic space available. Some transplanted retinas were, in fact, larger than both host retinas perhaps because of inherent genetic differences. Data on normal animals is from Reh and Constantine-Paton (1983). Data from three-eyed frogs is from Constantine-Paton and Ferrari-Eastman (1985).

----

[1]This analysis also revealed a volume increase of approximately 25% in the deeper tectal layers containing predominantly the somata of tectal neurons. Since there were no detectable changes in soma packing in the major layer where density could be measured, the volume increase probably indicates a real increase in tectal cells with double innervation.

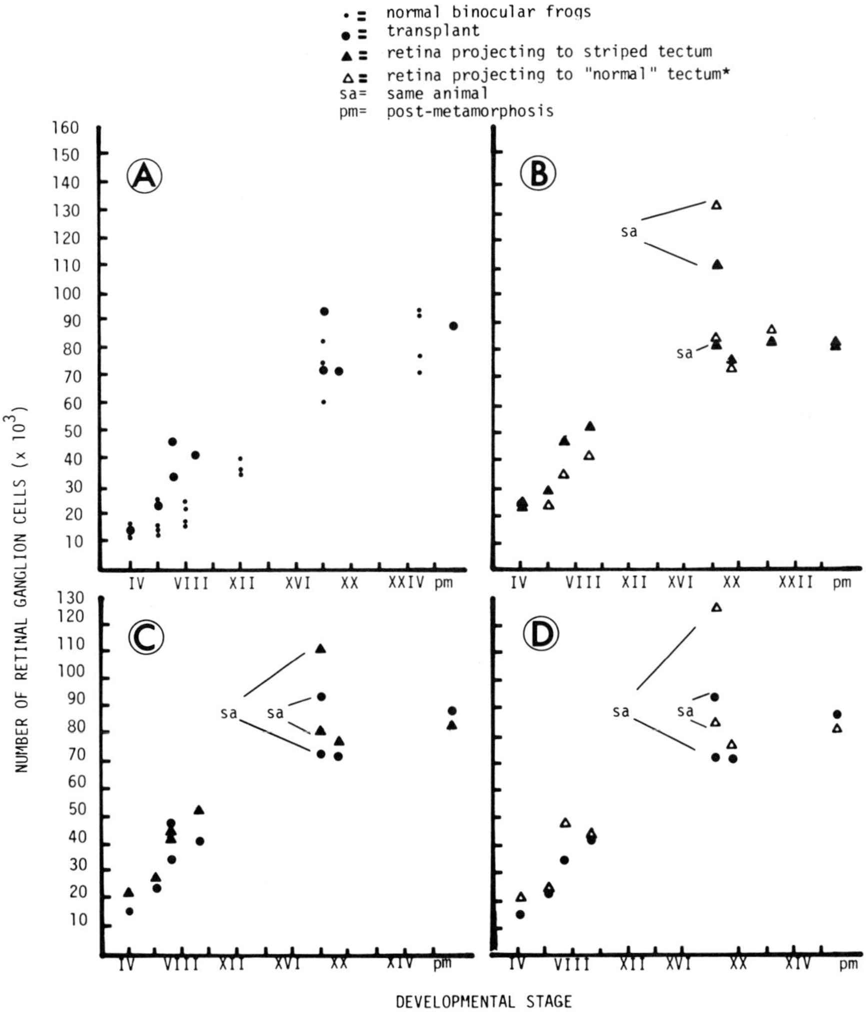

• = normal binocular frogs
● = transplant
▲ = retina projecting to striped tectum
△ = retina projecting to "normal" tectum*
sa= same animal
pm= post-metamorphosis
A
B
C
D
sa
NUMBER OF RETINAL GANGLION CELLS (x 10^3)
DEVELOPMENTAL STAGE
IV
VIII
XII
XVI
XX
XXIV
XXII
XIV
pm

tectum, the number of synapses per unit volume in doubly innervated tecta (N_V = 0.290) is approximately equal to the number of synaptic contacts per unit volume in singly innervated tecta (N_V = 0.310) of 3-eyed frogs. Furthermore, the number of synapses made within the singly and doubly innervated tectal lobes of three-eyed frogs is comparable to the number of synapses identified in the tectal lobes of a control frog of the same developmental stage. In short, the number of synapses in the superficial tectal neuropil appears to be a characteristic set by the optic tectum and not by the number of innervating retinal ganglion cells (Norden and Constantine-Paton, 1982, 1983).

The fine structural analysis has also demonstrated another characteristic of striped tectal lobes that provides some additional insight on synapse number regulation in this experimental system. The essential observation can be seen in figure 3. The photomicrographs in the left half of the figure are from a singly innervated tectum; the photographs on the right half are from the analogous region of the doubly innervated tectum from the same animal. The doubly innervated tectum contains much more extracellular space than the equivalent region of the singly innervated tectum. This difference between double and normally innervated tecta has now been observed in three-eyed animals of widely differing ages and it is completely counterintuitive. The retinal ganglion cell counts discussed above indicated, if anything, a significant increase in afferent axon number. Furthermore, light microscope measurments of the tangential extent of RGC terminal arborizations have failed to reveal any significant differences between singly and doubly innervated tecta (Reh and Constantine-Paton, 1984). The terminals also appear to be of the same caliber and arborization density under the two conditions, even though the EM analysis indicates that RGCs from each eye are making fewer total synaptic contacts. Thus simple volume occupied by afferents in striped tecta should be roughly twice that of the normal tectal lobe. Nevertheless, this volume is packed into a neuropil zone that shows only a 37% increase, at the light microscope level, in total volume when compared to the normal tectal lobe. If, as the EM analysis indicates, a considerably larger proportion of neuropil volume in doubly innervated tecta is extracellular space, then some cellular element(s) in the striped tectal neuropil must be reduced in volume.

TABLE 1

Morphometric analyses of synaptic densities in the singly (S) and doubly (D) innervated tectal lobes of two three-eyed frogs and a comparison with a normal animal of the same developmental stage. R = right; L = left.

PARAMETERS	3-eyes		Normal	
	S	D	R	L
1. Number of Synapses				
Total Number of Synapses Counted	2,479	2,673	823	956
Total Area Examined (μm^2)	26,460	42,336	8,820	10,584
Number per Unit Area (N_a)	0.094	0.063	0.090	0.090
2. Mean Synaptic Diameter ($\underline{d}$), in μm	0.300	0.278	0.280	0.290
3. Relative Tectal Volume	1	1.30	1	1
4. Number per Unit Volume ($N_v/\underline{d}$)				
Before Volume Correction	0.310	0.230	0.320	0.310
After Volume Correction	0.310	0.290	0.320	0.310

It seems likely that a partial solution to this apparent paradox will reside in a significant decrease in tectal neuron dendrites. Preliminary EM morphometric analyses suggest that a smaller percentage of neuropil area may be occupied by dendritic profiles in doubly as opposed to singly innervated optic tecta. We do not yet know if this reflects a decrease in tangential extent or density of some or all dendritic arbors. It could also reflect a loss of some distinct tectal neuron types. There are a number of very large neurons in normal frog tecta whose dendrites would certainly cover the area of more than one stripe (Potter, 1969). Perhaps cells such as these cannot maintain uncorrelated inputs from the two retinas and either die or trim their dendrites so that they only arborize in the terminal zones of one eye. Studies comparing the light microscope morphology of tectal neurons under single and double innervated conditions will hopefully resolve this issue.

EVIDENCE FOR INHERENT TECTAL MODULES

The idea that a particular cell type or some large multicellular circuit in the tectum must function as an indivisible functional unit that cannot receive input from both the host and the

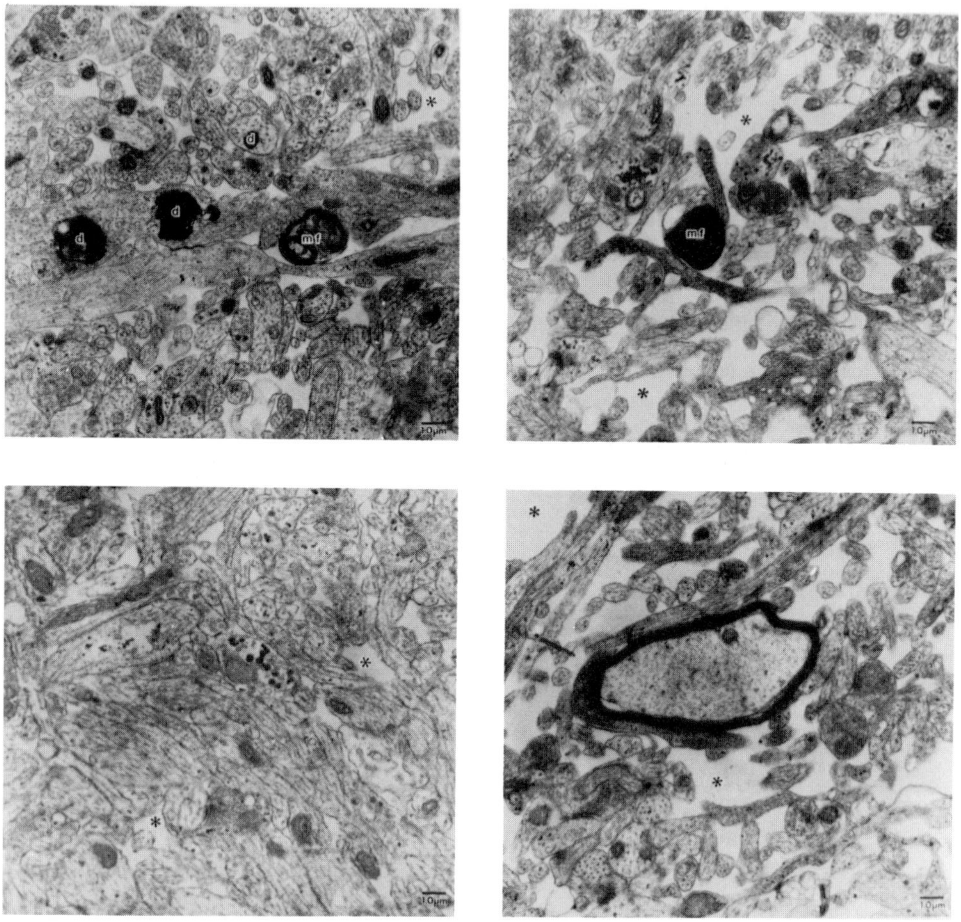

Figure 3. Fine structural effects of double innervation. Electron micrographs of neuropil from equivalent midtectal regions of a singly and a doubly innervated tectum of the same three-eyed frog sacrificed one month past metamorphosis. Left-hand side: micrographs of the singly innervated tectal neuropil. There are relatively small areas of extracellular space(*). A few myelin figures (MF) and some degeneration debris (d). This tectal neuropil was qualitatively the same as that seen in both tectal lobes of a normal binocular Rana pipiens at one month past metamorphosis. Right-hand side: micrographs of the doubly innervated tectal neuropil. The processing of this tectal lobe was identical to the singly innervated lobe, nevertheless, the lobe shows a striking increase in the amount of extracellular space.

TABLE 2

Average tectal neuropil volume occupied per retinal ganglion cell for normal and co-innervating eyes of three-eyed frogs.

Cubic microns/retinal ganglion cell[1]

Animal	Developmental Age	Normal Host Retina	Double Innervating Host Retina	Supernumerary Retina
490	T&K IV[2]	513	343	326
491	T&K VI	390	238	103
480	T&K VIII	484	417	238
476	T&K X	787	695	913
484	T&K XI	854	698	615
H90	T&K XVIII	1,089	413	989
551	1 week p.m.[3]	3,064	2,650	1,554
513	2 weeks p.m.	1,896	1,015	2,016
562	3 weeks p.m.	3,668	1,721	1,434
514	3 1/2 weeks p.m.	2,789	2,024	3,846
570	2 months p.m.	2,304	1,753	3,846

[1]Retinal ganglion cell number = Large total clear neuronal nuclei in the retinal ganglion cell layer after Konigsmark (1970) correction for double counts.

[2]Tadpole stages according to Taylor and Kollros (1946).

[3]p.m. = Post-metamorphosis. Metamorphosis is defined as the day the animal becomes air breathing.

supernumerary retina is also effective in explaining several other perplexing aspects of the striped termination pattern. For example,the light microscope morphometric data allows us to calculate the average neuropil volume occupied per ganglion cell for all three eyes of three-eyed frogs (Constantine-Paton and Ferrari-Eastman, 1985). As can be seen in the data presented in table 2, this value varies widely even within the same animal. Thus stripe volume does not appear to be directly related to innervation density. The numbers suggest instead that at some critical density (which may vary in animals of different developmental ages) axons from adjacent regions in one retina can take over a region of neuropil in an all or none fashion. Such a phenomenon is fully consistent with the kind of positive feedback or reinforcement of initial synapses from one eye that might be expected if cooperative activation of tectal neurons is involved in the activity-dependent sorting phase of map formation and eye-specific segregation. In such a situation the first synapses to temporally summate their activity and drive a tectal neuron could effectively recruit that tectal neuron into the cooperative network so that the domain of activation of the first spiking tectal cell would become the favored domain for subsequent synapses from the same retina. This speculation is also consistent with the highly conserved periodicity of the eye-specific stripes and with their consistent rostrolateral to caudomedial orientation. We would like to invoke the following corollary. The mechanism responsible for stabilizing synapses when their activity is relatively synchronized applies equally well to all sets of synapses that converge within the tectal lobe. Thus the position of synapses made by tectal neurons onto other tectal neurons should be capable of changing so as to maintain contacts within a similarly activated zone once two unrelated sets of retinal afferents make their initial synapses in any tectal region. However, the degree of this post- synaptic plasticity is likely to be limited by the morphological and neurochemical phenotypes of the tectal neurons and these phenotypes are probably set up independently by cellular interactions or cell lineages in the differentiating tectal neuroepithelium. In other words, eye-specific strips essentially crystallize around the smallest set of tectal neurons capable as a result of their independently determined phenotypes, of cooperative functional interactions. This line of reasoning implies that the global rostrocaudal orientation and the periodicity of the striped pattern reflects the inherent tangential orientation and the spacing of the facilitative circuits.

SUMMARY

Many of the ideas presented above are speculative. Nevertheless, the quantitative light and EM studies of three-eyed frogs establish that it is the optic tectum and not the retinas that limit retinotectal synapse number and that the area occupied by a striped projection is not proportional to the number of innervating ganglion cells. In light of these observations it does not seem unreasonable to assign to the tectum a determinative role in the spatial organization of the retinotectal synapses. The latter possibility suggests that the orientation and spacing of experimentally produced stripes may be revealing the size and disposition of the most basic functional domains dictated by neurogenesis in the optic lobe. Thus, comparisons of the cellular morphologies and functional circuits in doubly innervated tecta with those arising in normally innervated tecta may allow us to distinguish the true extent to which afferents to this region of the central nervous system are patterned by the circuits they activate. We should also be able to distinguish the extent to which afferents model the morphology and function of their post-synaptic tectal cells.

ACKNOWLEDGEMENTS

This work was supported by HHW grants EY01872 to MCP and EY03718 to JJN.

REFERENCES

Arnett, D. W. 1978. Statistical dependence between neighboring retinal ganglion cells in goldfish Exp. Brain Res. 32:49-53

Arnett, D. and T. E. Spraker. 1981. Cross correlation analysis of the maintained discharge of rabbit retinal ganglion cells. J. Physiol. Lond. 317: 29-47.

Changeux, J.-P. and A. Danchin. 1976. Selective stabilization of developing synapses as a mechanism for the specification of neuronal networks. Nature 264: 705-712.

Constantine-Paton, M. and R. R. Capranica. 1976. Axonal guidance of developing optic nerves in the frog. I. Anatomy of the projection from transplanted eye primordia. J. Comp. Neurol. 170: 17-32.

Constantine-Paton, M. and M. I. Law, 1978. Eye-specific termination bands in tecta of three-eyed frogs. Science 202: 639-641.

Constantine-Paton, M. and P. Ferrari-Eastman. 1985. Quantitative pre and post-synaptic correlates of intraocular competition and segregation in the frog. (Submitted for publication).

Constantine-Paton, M., E. C. Pitts and T. A. Reh. 1983. The relationship between retinal axon ingrowth, terminal morphology, and terminal patterning in the optic tectum of the frog. J. Comp. Neurol. 218: 297-313.

Cowan, W. M., J. W. Fawcett, D. D. M. O'Leary and B. B. Stanfield. 1984. Regressive events in neurogenesis. Sience 225: 1258-1265.

Cunningham, T. J. 1982. Naturally occuring neuron death and its regulation by developing neural pathways. Int. Rev. Cytol. 74: 163-186.

Fawcett, J. W. and D. J. Willshaw. 1982. Compound eyes project stripes on the optic tectum in Xenopus. Nature 296: 350-352.

Hubel, D. H. and T. N. Wiesel. 1965. Binocular interaction in striate cortex of kittens rared with artificial squint. J Neurophysiol. 28: 1041-1059.

Ide, C. F., S. E. Fraser and R. L. Meyer. 1983. Eye dominance columns from an isogenic double-nasal frog eye. Science 221: 293-295.

Konigsmark, B. W. 1970. Methods for the counting of neurons. In: Contemporary Research Methods in Neuroanatomy, W. J. H. Nauta and S. O. E. Ebbson (eds.), Springer-Verlag, New York, pp. 315-340.

Law, M. I. and M. Constantine-Paton. 1980. Right and left eye bands in frogs with unilateral tectal ablations. Proc. Natl. Acad. Sci. USA 77: 2314-2318.
Law, M. I. and M. Constantine-Paton. 1981a. Anatomy and physiology of experimentally produced striped tecta, J. Neurosci. 1: 741-759.
Law, M. I. and M. Constantine-Paton. 1981b. Morphometric examination of competing retinal projections in three-eyed frogs. Proc. Soc. Neurosci. 7: 405.
LeVay, S., D. H. Hubel and T. N. Weisel. 1975. The pattern of ocular dominance columns in macaque visual cortex revealed by a reduced silver stain. J. Comp. Neurol. 159: 559-576.
Mastronarde, D. N. 1983. Correlated firing of cat retinal ganglion cells. I. Spontaneously active inputs to X- and Y-cells. J. Neurphysiol. 49: 303-324.
Meyer, R. L. 1982. Tetrodotoxin blocks the formation of ocular dominance columns in goldfish. Science 218: 589-591.
Meyer, R. L. 1983. Tetrodotoxin inhibits the formation of refined retinotopography in goldfish. Dev. Brain Res. 6: 293-298.
Norden, J. J. and M. Constantine-Paton. 1982. Evidence for post-synaptic regulation of synaptic density in three-eyed frogs. Proc. Soc. Neurosci. 8: 450.
Norden, J. J. and M. Constantine-Paton. 1983. Segregation and density of synapses in eye-specific termination bands in the tecta of 3-eyed frogs. Proc. Soc. Neurosci. 9: 59.
Pollack, J. G. and T. L. Kickey. 1979. The distribution of retino-collicular terminals in rhesus monkey. J. Comp. Neurol. 185: 587-602.
Potter, H. D. 1969. Structural characteristics of cell and fiber populations in the optic tectum of the frog (Rana catesbeiana). J. Comp. Neurol. 136: 203-232.
Reh, T. A. and M. Constantine-Paton. 1983. Qualtitative and quantitative measures of plasticity during the normal development of the Rana pipiens retinotectal projection. Dev. Brain Res. 10: 187-200.
Reh, T. A. and M. Constantine-Paton. 1985. Eye-specific segretation requires neural activity in three-eyed Rana pipiens. J. Neurosci. 5:1132-1143.
Reh, T. A., E. Pitts and M. Constantine-Paton. 1983. The organization of the fibers in the optic nerve of normal and tectum-less Rana pipiens. J. Comp. Neurol. 218: 282-296.
Schmidt, J. T. an dD. L. Edwards. 1983. Activity sharpens the map during the regeneration of the retinotectal projection in goldfish. Brain Res. 269: 29-39.
Stent, G. S. 1973. A physiological mechanism for Hebb's postulate of learning. Proc. Natl. Acad. Sci. USA 70: 997-1001.
Taylor, A. C. and J. J. Kollros. 1946. Stages in the development of Rana pipiens Larvae. Anat. Rec. 94: 7-23.

The Development of the Map from the Nucleus Isthmi: The Influence of Visual Experience on the Formation of Orderly Connections in the Visual System

Susan Boymel Udin

The high degree of anatomical complexity of organization in the visual system is one of its most notable features. Many variations are played on the theme of retinotopic mapping--distortions, transformations and subdivision of maps have been extensively documented in a wide range of vertebrates (Allman and Kaas, 1974; Hubel and Wiesel, 1974; Gruberg and Udin, 1978; Grobstein and Comer, 1983). One feature which most of the visual maps share in common is that the representations from the two eyes are in register. For example, a binocular cell in striate cortex has identical or nearly identical receptive field locations from both eyes. What developmental processes establish this precision of matching of input from the two eyes? This question has received intense attention from many investigators, prompted in large part by the pioneering work of Hubel and Wiesel, who demonstrated that the establishment of binocular connections can be disrupted by monocular visual deprivation or by strabismus (Wiesel and Hubel, 1965; LeVay, et al., 1980). Such manipulations prevent individual cells from maintaining responsiveness to both eyes, but the overall topography of the ipsilateral and contralateral geniculocortical maps is not disrupted. However, subsequently it has been found that visual deprivation and strabismus do disrupt one of the other topographic cortical projections, namely, the callosal fiber system which normally connects the representations of the vertical meridian in the two hemispheres (Choudhury et al., 1965). In the absence of normal visual experience, the topography of this system develops aberrantly, with some fibers arising from improperly located cells and some fibers terminating in improper locations (Innocenti and Frost, 1979; Lund and Mitchell, 1979; Berman and Payne, 1983;

Elberger et al., 1983; von Grunau and Rauschecker, 1983). Important and intriguing as the callosal projection is, it is not a simple system to study. For example, callosal cell bodies are intermingled with many other types of cell bodies. In addition, the initial development of callosal projections takes place prenatally (Innocenti and Caminiti, 1980).

Another system is available, however, in which we may study some of the properties of a topographic linkage which is highly influenced by early visual input. This system is the isthmotectal projection of the South African claw-toed frog, Xenopus laevis. The isthmotectal pathway links the two optic tecta, the largest structures of the amphibian central visual system (Gruberg and Udin, 1978). Each tectum receives a direct projection from its contralateral retina; however, the ipsilateral eye's input to the tectum is indirect. Ipsilateral information is relayed by the tecto-isthmo-tectal pathway illustrated in figure 1. When this linkage develops correctly, the nucleus isthmi provides the tectum with a map of the ipsilateral eye's visual field which is in register with the retinotectal map from the contralateral eye (Gaze and Jacobson, 1962; Keating, 1974).

The projection from the nucleus isthmi is of special interest because it exhibits extraordinary plasticity during development. Visual deprivation prevents the formation of an orderly ipsilateral map by the isthmotectal projection (Keating and Feldman, 1975). Moreover, any change in the relative positions of the two eyes can induce an abnormal topography of the ipsilateral map. For example, as detailed below, a manipulation such as rotating one eye 180° within its orbit can cause such extreme changes in the trajectories of isthmotectal axons that axons which are normally destined for one pole of the tectum will be rerouted to the opposite pole (Udin and Keating, 1981; Udin, 1983).

In addition to the malleability of its connections, the nucleus isthmi of Xenopus offers other experimental advantages, such as its physical isolation from the rest of the visual system, the restricted number of its inputs and targets, and its accessibility at all stages of development (Udin and Fisher, 1985).

In this paper, I will describe the physiological and morphological consequences of abnormal visual experience in Xenopus, and I will discuss the role of plasticity in the normal development of binocular vision in Xenopus. The central hypothesis which will

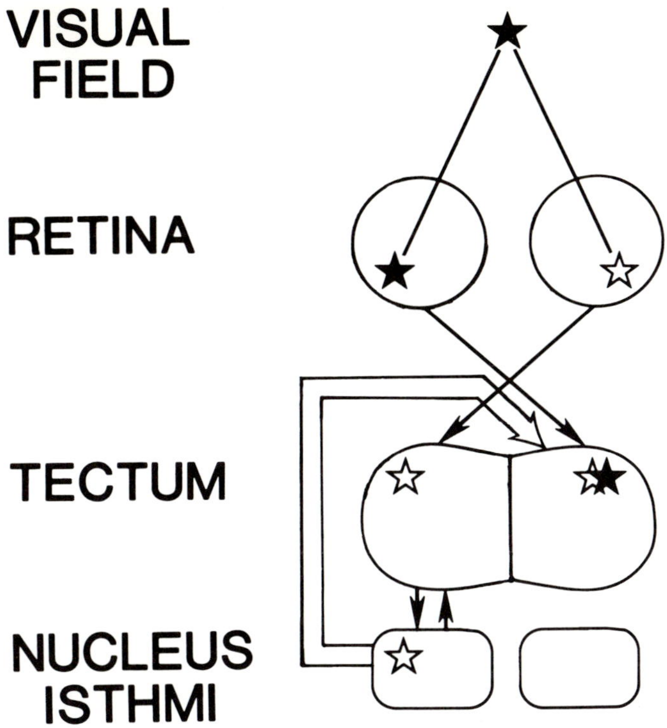

Figure 1: Binocular input converges on both tecta. This diagram illustrates the routes taken by fibers relaying input to the right tectum. The star represents a location in the binocular visual field. The left eye projects directly to the right tectal lobe. The right eye's projection is relayed first to the left tectum and then to the left NI. Axons from the left NI cross the midline in the postoptic commissure en route to the right tectum. The left NI also projects to the left tectum. Binocular input converges on the left tectum by a route symmetrical to that illustrated here.

be invoked to help explain these phenomena is a modification of Hebb's model (Hebb, 1949; Keating, 1974): it is proposed that isthmotectal axons use visually-evoked activity patterns to determine where to arborize. An isthmic terminal will be stabilized only if it encounters a region of the tectum in which the retinotectal axons have similar receptive fields and therefore similar activity patterns. In support of this hypothesis, I will show that developing isthmotectal axons branch widely and seem to "explore" various tectal regions in order to find neighboring arbors with matching activity. In addition, I will discuss the role which non-visual mechanisms play in establishing a crude preliminary order in early development.

THE PROBLEM OF VARIATION IN EYE POSITION DURING DEVELOPMENT

One of the general functions of plasticity during development may be to allow for fine tuning of connections to help compensate for minor strabismus, variations in interocular separation, cyclotorsion, etc., especially as those features change during early life. All of these variations can potentially influence the topographic matching of the inputs from the two eyes.

In Xenopus, the establishment of congruent maps from the two eyes poses a special challenge because of the peculiar development of that species: Xenopus' eyes undergo a remarkable change in position as the animals grow (Grobstein and Comer, 1977; Grant, 1982) (figure 2). In young tadpoles, the eyes face laterally and have essentially no binocular overlap. During metamorphosis, the eyes begin to shift to a fronto-dorsal position, and this movement eventually establishes a binocular overlap of 170° in adults. The increase in binocularity is accompanied by a readjustment of the isthmotectal projection, which brings input to the tectum from the ipsilateral eye. Initially, isthmotectal arbors have no proper targets in the tectum because the fields which the axons relay from the ipsilateral eye do not correspond to any of the fields relayed from the contralateral eye. As the eyes begin to move, however, a binocular zone is created in the visual field above the head at about stage 60. (See figure 2.) Thus, the ipsilateral fields of some of the isthmotectal axons now overlap with the contralateral fields of some of the retinotectal axons. Those retinotectal axons which first subserve a binocular region of space terminate at the very front of the tectum. At the same time, the first electrophysiologically-detectable ipsilateral units are recorded--in register with the contralateral units--in this same small portion of the tectum (Beazely et al., 1972; Grant, 1982). As the eyes continue to move, more and more of the retinotectal projection subserves the binocular portion of the field, and ipsilateral units

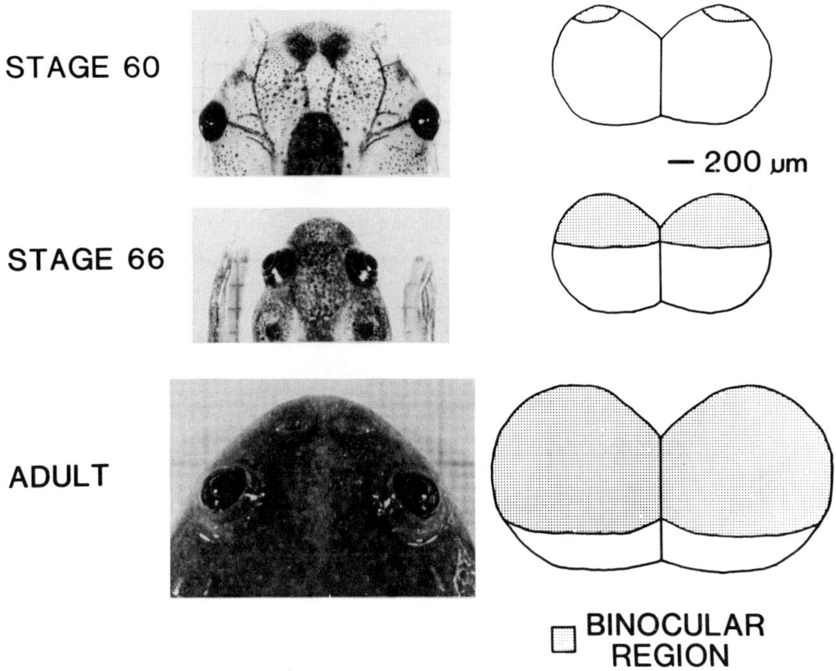

Figure 2: Photographs show tadpoles (stages 60 and 66) and 3-year adult Xenopus. Stage 60 tadpole is transparent, and only the dorsal aspect of each eye is visible. Note the eye migration which takes place with age. Line drawings show changes in tectal size; stippling shows regions corresponding to binocular visual field; the remainder of each tectum receives retinotectal input corresponding to the monocular "crescent". (Redrawn from Grant, '82).

can be recorded throughout the expanding binocular region of the tectum. As will be discussed below, this establishment of an orderly ipsilateral map is contingent on normal visual input.

THE NORMAL TECTO-ISTHMO-TECTAL RELAY

In order to understand the anatomical changes in isthmotectal connectivity which underlie the changes in the ipsilateral map, it was first necessary to work out the basic circuitry of the isthmotectal system in normal adults. Figure 1 shows that the right tectal lobe receives input directly from the left eye via the retinotectal projection and indirectly from the right eye via the tecto-isthmo-tectal relay. Each of these projections is topographic, and under normal circumstances, the projections from the two eyes to each tectal lobe are in register (Grobstein et al., 1978; Gruberg and Udin, 1978; Grobstein and Comer, 1983).

The details of the anatomical connections from the tectum to the NI and from the NI to the tecta in Xenopus were determined using horseradish peroxidase (HRP) (Udin & Keating, 1981). Localized injections of HRP were made in the tectum. By examining the anterograde transport of HRP from the tectum to the NI, the topography of the tecto-isthmic projection was determined (figure 3, right). Next, the pattern of retrograde labeling in the NI contralateral to the injected tectum was established (figure 3, left). Combining these sets of results demonstrated that information about a particular region of the visual field is relayed from one tectum to the proper position in the other tectum via the nucleus isthmi.

More recently, HRP has been used to reveal the trajectories and branching patterns of individual isthmotectal axons. HRP is injected into the NI and is transported anterogradely to the opposite tectum (Udin, 1983). The tectum is removed, flattened, and reacted whole, so that individual axons can be traced from the caudal diencephalon to their terminal arbors in the tectum. In other cases, the tectum is embedded and sectioned in order to study lamination patterns.

Analysis of filled axons reveals that there are two laminae of axons. The superficial axons travel in the stratum zonale (layer A of Potter (1969)) and make compact terminal arbors within that layer (see figure 4). The great majority of axons run in a rostrocaudal direction from the point at which they enter the tectum to the position at which they arborize. This pattern implies that each axon somehow "chooses" the correct position along the mediolateral dimension of the tectal border; once having chosen the appropriate position, it need only be instructed about how far caudally to make its arbor. I will refer to these patterns below, where they will be compared to the trajectories and arborizations of superficial axons in abnormally-reared adults and in developing normal frogs.

The other lamina of isthmotectal axons and arbors is located in layer 8 and the deeper half of layer 9. The axons in this lamina also travel in a rostrocaudal direction. Their arborizations are quite heterogeneous, ranging from quite compact to extremely large (up to 500 μ long in a tectum of 1500 μ total length). (See figure 5.) These deep axons will not be discussed further.

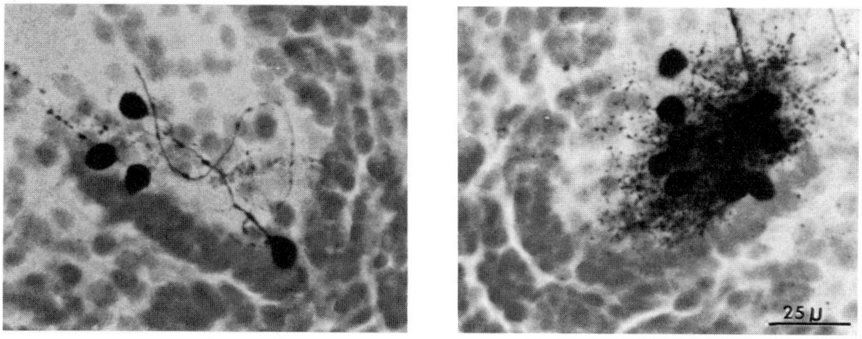

Figure 3: Horizontal sections through the left and right NI after an HRP injection in the right tectal lobe in a tadpole. The left side (left NI) shows retrogradely filled cell bodies and their axons. The right side (right NI) shows both retrogradely filled cell bodies and anterogradely filled tecto-isthmic axons. Rostral is upward.

THE CONSEQUENCES OF ABNORMAL VISUAL INPUT DURING DEVELOPMENT

Keating and his collaborators (Gaze et al., 1970; Keating, 1974) were the first to ask what mechanisms brought the frog's ipsilateral map into register with its contralateral map. In order to test the possible organizing role of binocular visual input, they reared Xenopus tadpoles with various sorts of abnormal visual experience. They demonstrated that dark-rearing prevents ipsilateral units from coming into topographic register with contralateral units. Although the topographic order of the ipsilateral map in a dark-reared animal is relatively normal with respect to the mediolateral tectal axis, it is profoundly disordered along the rostrocaudal axis (Keating and Feldman, 1975; Grant, 1982). Thus, binocular visual deprivation prevents the ipsilateral map from matching properly with the contralateral map.

In addition, Keating and his associates were able to induce formation of an abnormal - but orderly - ipsilateral map, by rotating one eye in mid-larval tadpoles (Gaze et al., 1970; Keating, 1975). Despite the fact that one eye is upside down while the other is right side up, each ipsilateral map winds up orderly and in

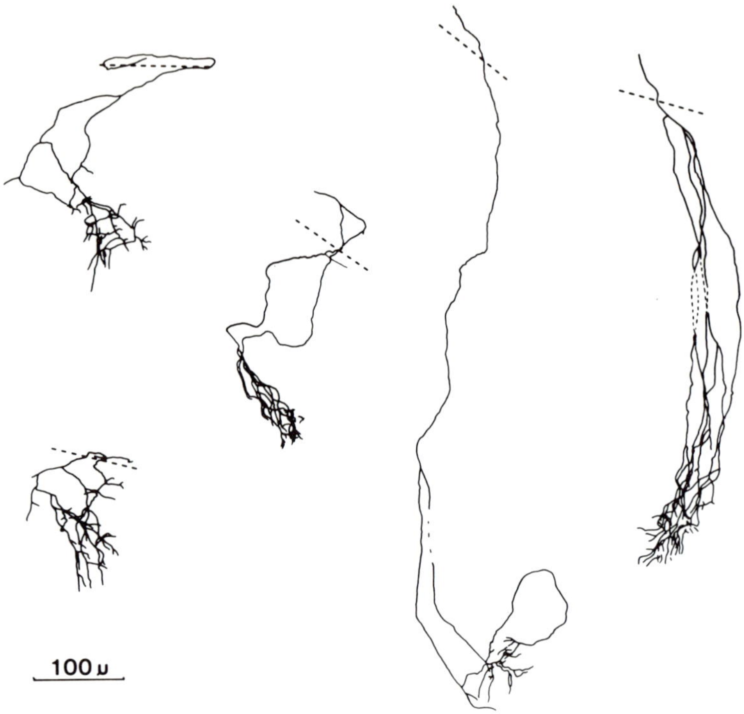

Figure 4: Camera lucida drawings of superficial axons in Xenopus of 4-9 months post-metamorphosis. The position at which each axon enters the tectum from the diencephalon is indicated by the dashed line. Caudal is downward.

register with the contralateral map to the same tectal lobe. (The rearrangement does not occur in eye-rotated animals which are dark-reared.) The anatomical basis of this rearrangement is shown diagrammatically in figure 6. The left column shows that rotating the right eye produces a rotated retinotectal map on the left tectum. Note that each ganglion cell still projects to its normal tectal target after eye rotation, but the cells now relay different visual information from normal. The middle of figure 6 shows that the retino-tecto-isthmic relay from the unrotated left eye remains normally oriented; it sends normally oriented maps to the right tectum and NI in both sets of animals. The right hand column of

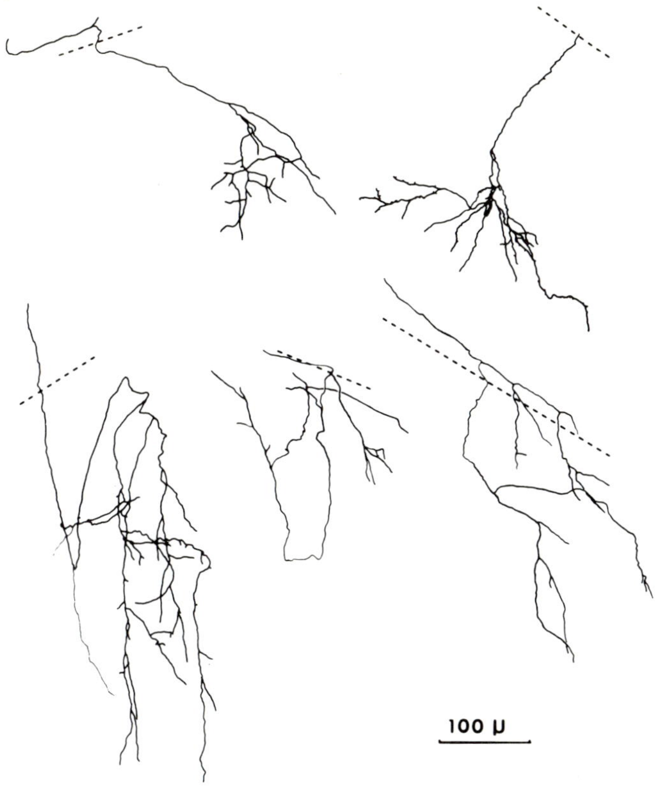

Figure 5: Deep axons in 4-9 month post-metamorphic Xenopus. Scale bar: 100 μm.

figure 6 shows how the projection from the right NI, with its normal map, changes when confronted by a rotated map on the left tectum: the axons change routes and grow to locations which have matching retinotectal receptive fields (Udin and Keating, 1981; Udin, 1983). For example, the isthmic axon with visual field B should terminate near a retinotectal axon which also has its receptive field at visual field position B. Normally, then, isthmic axon B terminates caudolaterally, but eye rotation makes it relocate to rostromedial tectum. Not shown in figure 6 are the corresponding changes which take place in the projection from the left NI, which receives a rotated map but projects to the tectum with normal retinotectal input.

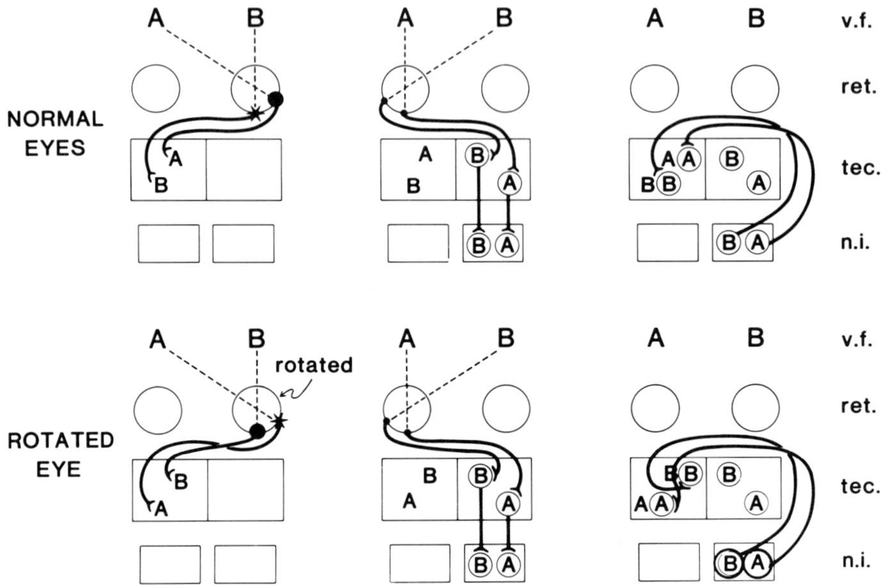

Figure 6: Schematic comparison of visual pathways in normal Xenopus (top row) with pathways in Xenopus reared with a 180° rotation of the right eye (bottom row). Left column: retinotectal projections from right eye. The star and filled circle illustrate two ganglion cells whose positions are interchanged as a result of the eye rotation. Middle column: retinotectal connections from the left eye are identical in both normal and eye-rotated animals, as are the connections from the right lobe of the tectum to the right NI. Right column: isthmotectal axons from the right NI follow normal trajectories toward the tectum, but eye-rotation induces them to follow abnormal routes within the tectum.

The evidence that isthmotectal axons take abnormal trajectories in eye-rotated Xenopus comes from studies using horseradish peroxidase (HRP) to trace connections. Just as we used localized tectal injections of HRP to establish the normal topography of tecto-isthmo-tectal connections, we employed the same techniques to study topography after early eye rotation (Udin & Keating, 1981). We used our knowledge of the normal connections plus our electrophysiological maps to predict how the isthmotectal topography would differ from normal. For example, after rotation of one eye by 180°, an isthmotectal axon which normally projects to rostral tectum should move to a caudal position, and an axon which normally projects to caudal tectum should be rerouted to rostral tectum. Such a result is shown in figure 7. Cells in the dorsomedial part

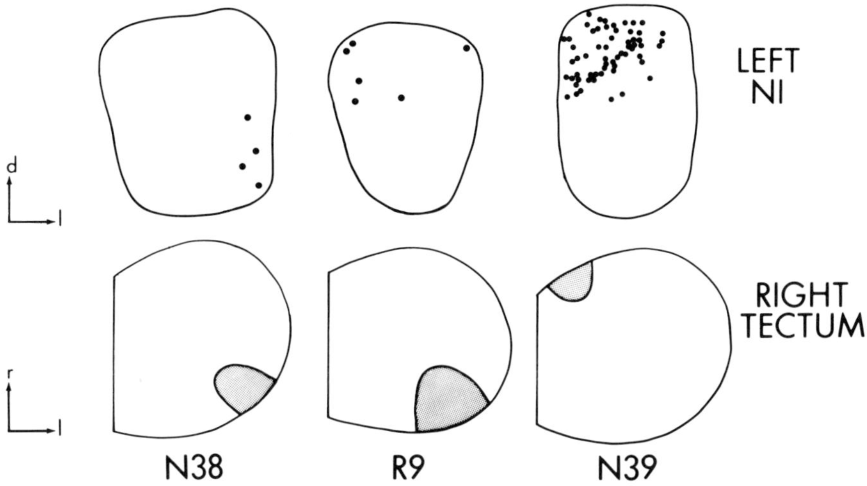

Figure 7: The crossed isthmotectal projection in Xenopus R9 is compared with two matched controls. In the lower part of the figure, the left optic tecta of the three animals are shown in dorsal view. The sites of the localized HRP injections are indicated by the filled regions. In the upper part of the figure, the right NI of the three animals are shown in rostral view. The positions within the NI of cells labeled by the tectal injection of HRP are indicated by black dots. Reprinted from Udin and Keating (1981).

of the NI are normally labelled after an HRP injection in rostromedial tectum; in contrast, after early 180° eye rotation, these cells are labelled by injections in the opposite pole of the tectum. Therefore, cells which normally project to rostromedial tectum are rerouted to caudolateral tectum as a consequence of the abnormal visual input during development. In some other animals with eye-rotations, though, the labeling pattern is not so easily interpreted. Rather than finding that one tectal injection labelled just one group of cells in the NI, we often observed that a single HRP injection in eye-rotated frogs labelled two foci of cells in the NI. (See figure 8.) One of these groups was where we'd predicted it to be on the basis of the eye rotation, but the other group was

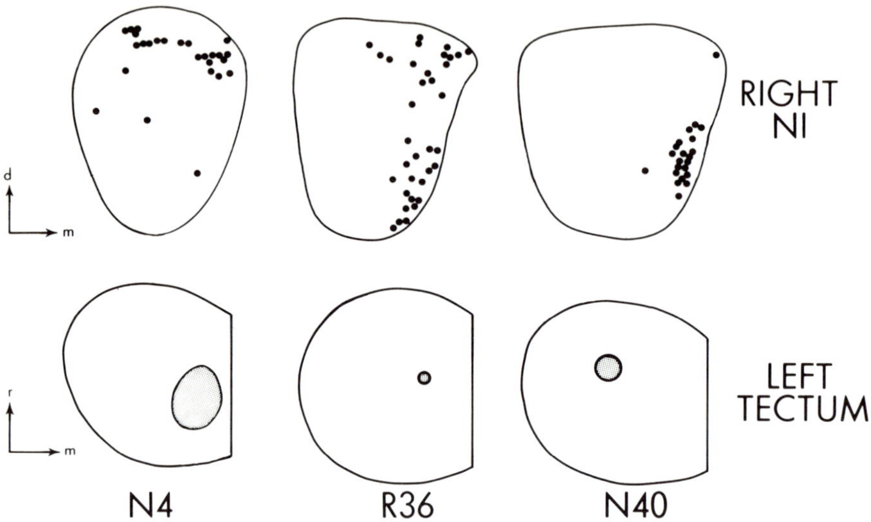

Figure 8: The crossed isthmotectal projection in animal R36 compared with two matched controls. Same conventions as in figure 7. Reprinted from Udin and Keating (1981).

located where we would have found labelled cells in a comparable control case. The simplest explanation seemed to be that axons were growing by circuitous routes, first approaching their "normal" targets and then turning to reach their new targets. A single injection of HRP therefore could label two groups of cells, one because the HRP was picked up by terminals of axons which ended at the injection site, and the other because HRP was taken up by damaged axons which passed through the injection site en route to new locations. For example, in figure 6, an injection in caudolateral tectum in an eye-rotated frog would label isthmic cell B as well as isthmic cell A.

Retrograde tracing data, then, suggested that eye rotation induces isthmo-tectal axons to take abnormal trajectories. I further investigated this possibility by anterograde labelling of the isthmo-tectal axons. I injected HRP into the NI and was then able to follow the routes of isthmic axons in the opposite tectum (Udin, 1983). As indicated above, each axon chooses a particular entry point at the tectal border; most axons enter the tectum at its rostral border, while others enter from the medial or lateral edges. Once the axons have entered the tectum, they normally grow straight caudally to reach their proper termination sites. (See figure 9A.) In contrast, many axons in eye-rotated animals take obviously abnormal routes within the tectum. They seem to enter at their normal positions but are then forced to criss-cross and make u-turns or even loops as they seek out appropriate places to terminate. (See figure 9B.) Their routes may take them through their normal target zones en route to the new sites. These results demonstrate that the trajectories of growing isthmic axons are profoundly influenced by early eye rotation.

THE MORPHOLOGY OF GROWING AXONS

The results described above suggest that the final morphology of isthmotectal axons could be the result of several different mechanisms. First, the axons are prompted to array themselves in a roughly topographic fashion across the mediolateral axis of the tectum. Then they grow caudally into the tectum and begin to arborize. Visual input determines which branches are retained and which are retracted, and abnormal visual input can even force them to abandon their rostrocaudal orientation.

In order to see when these different facets of development occur, I have labelled isthmotectal axons in developing tadpoles and juvenile Xenopus.

STAGES 57-60

The first question I wished to answer was whether isthmotectal axons can enter the tectum before the eyes have begun to move and to establish a zone of binocular overlap overhead. In other words, do binocular visual cues induce the axons to enter the tectum, or are other forces providing the impetus? The results favor the latter option, because many HRP-filled axons enter the tectum well before eye migration begins, and they can be found in virtually the entire tectum. (See figure 10.) They show a pronounced tendency to grow straight caudally. This behavior is consistent with their mature pathways. However, at these early stages, the axons bear few branches. Thus, prior to the onset of binocular vision, axons have established a mediolateral distribution and set down a rostrocaudal

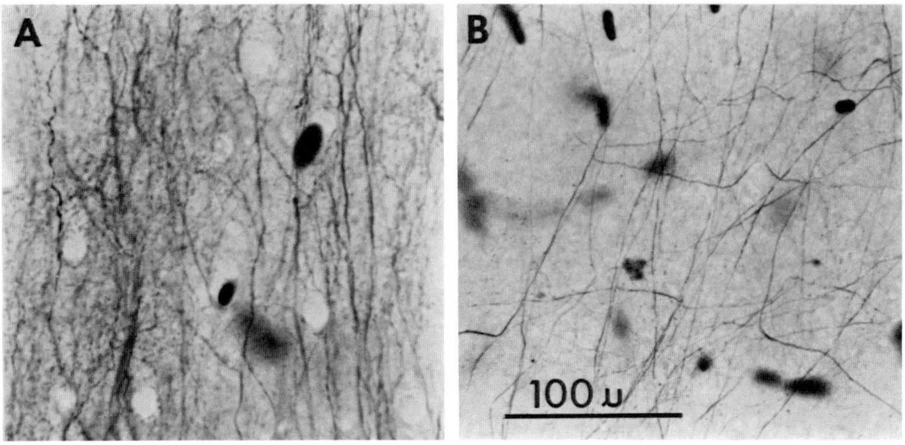

Figure 9: Isthmotectal axons in flat-mounted tecta of adult normal (A) and eye-rotated (B) Xenopus. Dark oval and round spots are red blood cells. Scale bar: 100μm. Caudal is downward.

scaffold upon which to establish branches.

STAGES 61 TO 2 WEEKS POST-METAMORPHOSIS

During the 3-4 weeks which elapse between the onset of eye migration at stage 60 and the end of metamorphosis at stage 66, the eyes move at a rapid rate and the amount of binocular overlap increases to about 100° (Grant, 1982). Throughout this period, despite the rapidly changing relationship between the two eyes, presynaptic activity of ipsilateral units can be recorded, and they remain in register with the contralateral units (Grant, 1982). These observations raise several questions. First, do the axons begin to produce more branches, as one would predict from the fact that their activity can now be recorded above noise level? As figure 11 shows, there is indeed a dramatic increase in the number of branches, and many axons now have relatively dense clusters of branches which probably constitute the electrically-detectable part of the arbor. Another question is whether the axons are confined to the currently-binocular region, or whether they also can extend branches into the still-monocular zone. Figure 11 demonstrates that isthmotectal axons can occupy both zones.

The axons still maintain their initial rostrocaudal orientation. This observation reinforces the idea that the axons'

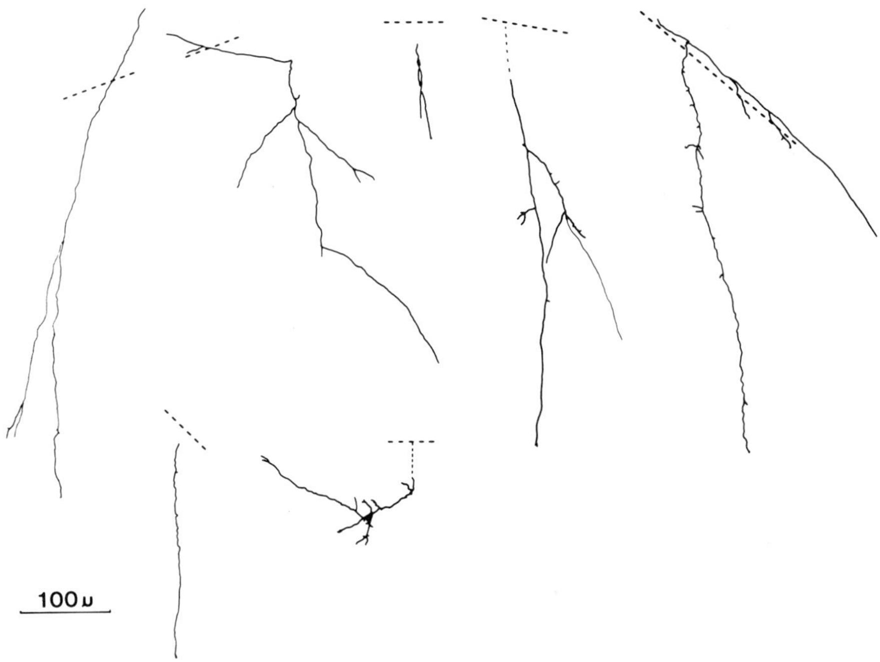

Figure 10: Superficial axons in tadpoles of stages 57-60.

initial mediolateral positions are generally correct and that they need only to elaborate branches at appropriate positions along their rostrocaudal extents.

The overall sizes of these immature arbors are about twice as large as mature arbors. In order to obtain a quantitative measure of arbor sizes, I have computed the mean inter-terminal distances (ITDs). (See figure 12.) This method yields a mean inter-terminal distance of 104 µm for superficial axons from animals from stage 65 to 2 weeks post-metamorphosis; for late juveniles (4-9 months post-metamorphosis), superficial axons have a mean ITD of 51 µm.

Why are the immature arbors so large? There are at least two possibilities. First, there may be an intrinsic tendency for branches to extend long distances during development; this would be consistent with the axons' tendency to grow far into the tectum even before eye migration begins. Alternatively, the large arbor size may be a reflection of the extraordinary demands of keeping pace

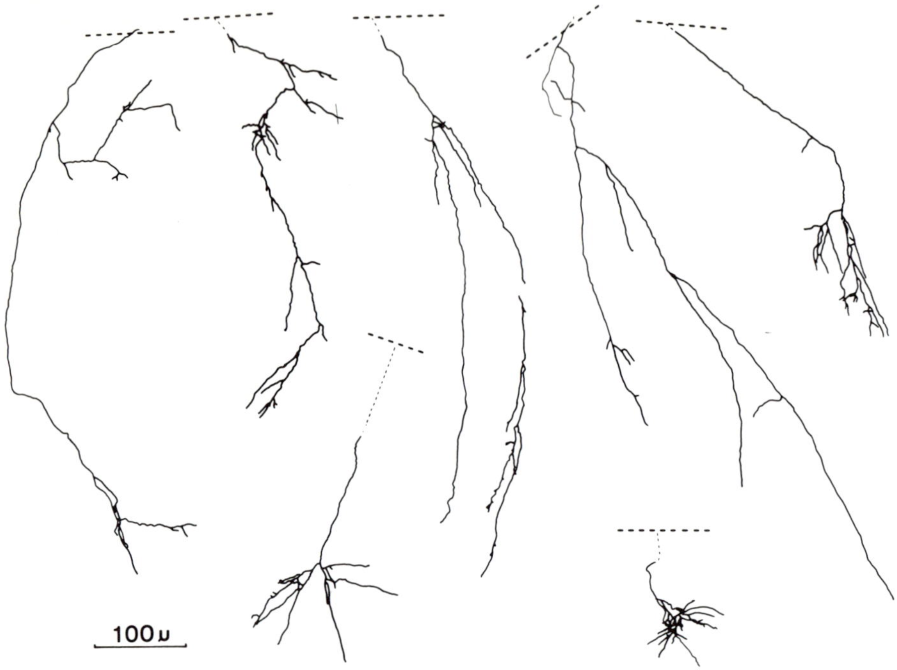

Figure 11: Superficial axons between stage 65 and two weeks post-metamorphosis.

with rapid eye movement. For example, an axon which should arborize at the rostral border of the tectum at stage 60 will have to shift its arbor to the middle of the tectum, 500 μm caudal, in the space of one or two weeks. This rate of movement may not allow sufficient time for inappropriate branches to be completely eliminated in favor of "correct" ones.

4-12 WEEKS POST-METAMORPHOSIS

During this period, eye migration slows considerably, and the binocular field only increases from 100° to about 110°. Thus, the demands for shifting are greatly reduced. However, the axons are still capable of dramatic rearrangements such as those induced by early eye rotation. This capacity for reorganization is demonstrated by the results of electrophysiological mapping of eye-rotated froglets. Surprisingly, after <u>pre</u>-metamorphic eye rotation,

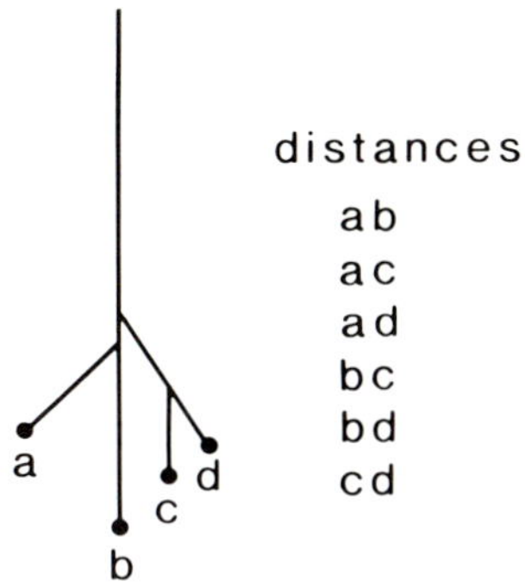

Figure 12: Mean interterminal distance is computed by measuring distances between all pairs of terminals and calculating the mean.

the maps recorded in the first few weeks after metamorphosis are essentially normal; only at about 1-3 months post-metamorphosis do the ipsilateral maps shift orientation to reflect the new relationship between the two eyes (Grant and Keating, 1985). Thus, the pattern of connections shown in the bottom half of figure 6 is not achieved until several weeks or months after the eye is rotated. These surprising results imply that the isthmotectal axons first arborize in approximately normal positions but then re-locate their branches to new positions. Is there a morphological feature which we can correlate with this ability of early post-metamorphic axons to establish new pathways? I would like to speculate that there are "test" branches produced by the axons which allow them to "explore" regions of the tectum at some distance from the main arbor. The arrows in figure 13 point to branches which may fulfill this function. Reference to figure 4 will show that these "test branches" largely disappear at later ages, when the axons have lost their plasticity.

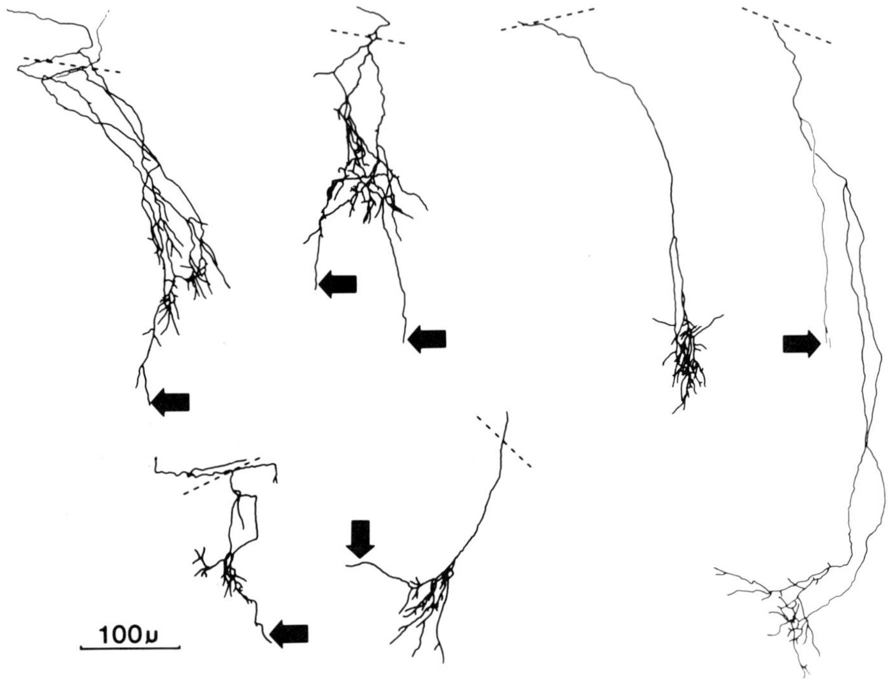

Figure 13: Superficial axons in 4-12 week post-metamorphic Xenopus. Axons point to "test branches."

CONCLUSIONS

The isthmotectal projection of Xenopus displays remarkable plasticity. During development, a rudimentary mediolateral order is established by non-visual mechanisms, and normal visual input provides the information which keeps the maps from the two eyes in register as the eyes change position; abnormal visual input can induce the establishment of an orderly but abnormal topography which brings the ipsilateral map into accord with an abnormal contralateral map.

ACKNOWLEDGMENTS

This work was supported by Public Health Service Grant EY-03470, by Basic Research Grant 1-884 from March of Dimes Birth Defects Foundation, and by Travel Grants from the Burroughs-Wellcome Trust.

REFERENCES

Allman, J.M. and J.H. Kaas. 1974. The organization of the second visual area (V II) in the owl monkey; a second order transformation of the visual hemifield. Brain Res. 76:247-265.

Beazley, L., M.J. Keating and R.M. Gaze. 1972. The appearance, during development, of responses in the optic tectum following visual stimulation of the ipsilateral eye in Xenopus laevis. Vis. Res. 12:407-410.

Berman, N. and B.R. Payne. 1983. Alterations in connections of the corpus callosum following convergent and divergent strabismus. Brain Res. 274:201-212.

Choudhury, B.B., D. Whitteridge and M.E. Wilson. 1965. The function of the callosal connections of the visual cortex. Q. J. exptl. Physiol. 50:214-219.

Elberger, A.J., E.L. Smith and J.M. White. 1983. Spatial dissociation of visual inputs alters the origin of the corpus callosum. Neurosci. Lett. 35:19-24.

Gaze, R.M. and M. Jacobson. 1962. The projection of the binocular visual field on the optic tecta of the frog. Quart. J. exp.Physiol. 47:273-280.

Grant, S. 1982. The development and modification of binocular neuronal connections in Xenopus laevis. Ph.D. Thesis, University of London, 405 pp.

Grant, S. and M.J. Keating. 1985. Time-course of changes in intertectal connections following eye-rotation in Xenopus laevis. In press.

Grobstein, P. and C. Comer. 1977. Postmetamorphic eye migration in Rana and Xenopus. Nature 269:54-56.

Grobstein, P. and C. Comer. 1983. The nucleus isthmi as an intertectal relay for the ipsilateral oculotectal projection in the frog, Rana pipiens. J. Comp. Neurol. 217:54-74.

Grobstein, P., C. Comer, H. Hollyday and S.M. Archer. 1978. A crossed isthmo-tectal projection in Rana pipiens and its involvement in the ipsilateral visuo-tectal projection. Brain Res. 156:117-123.

Gruberg, E.R. and S.B. Udin. 1978. Topographic projections between the nucleus isthmi and the tectum of the frog Rana pipiens. J. Comp. Neurol. 179:487-500.

Hebb, D.O. 1949. The organization of behavior. John Wiley and Sons, New York.

Hubel, D.H. and T.M. Wiesel. 1974. Uniformity of monkey striate cortex: A parallel relationship between field size, scatter, and magnification factor. J. Comp. Neurol. 158:295-306.

Innocenti, G.M. and R. Caminiti. 1980. Postnatal shaping of callosal connections from sensory areas. Exp. Brain Res. 38:381-394.

Innocenti, G.M. and D.O. Frost. 1979. Effects of visual experience on the maturation of the efferent system to the corpus callosum. Nature 280:231-233.

Keating, M.J. 1974. The role of visual function in the patterning of binocular visual connections. British Medical Bulletin 30:145-151.

Keating, M.J. and J. Feldman. 1975. Visual deprivation and intertectal neuronal connections in Xenopus laevis. Proc. R. Soc. Lond. B. 191:467-474.

LeVay, S., T.N. Wiesel and D.H. Hubel. 1980. The development of ocular dominance columns in normal and visually deprived monkeys. J. Comp. Neurol. 191:1-51.

Lund, R.D. and D.E. Mitchell. 1979. Asymmetry in the visual callosal connections of strabismic cats. Brain Res. 167:176-179.

Potter, H.D. 1969. Structural characteristics of cell and fiber populations in the optic tectum of the frog (Rana catesbeiana). J. Comp. Neurol. 136:203-232.

Udin, S.B. 1983. Abnormal visual input leads to development of abnormal axon trajectories in frogs. Nature 301:336-338.

Udin, S.B. and M.D. Fisher. 1985. The development of the nucleus isthmi in Xenopus laevis: I. Cell genesis and the formation of connections with the tectum. J. Comp. Neurol. 232:25-35.

Udin, S.B. and M.J. Keating. 1981. Plasticity in a central nervous pathway in Xenopus: anatomical changes in the isthmo-tectal projection after larval eye rotation. J. Comp. Neurol. 203:575-594.

Von Grunau, M.W. and J.P. Rauschecker. 1983. Natural strabismus in non-Siamese cats: lack of binocularity in the striate cortex. Exp. Brain Res. 52:307-310.

Wiesel, T.N. and D.H. Hubel. 1965. Binocular interaction in striate cortex of kittens reared with artificial squint. J. Neurophysiol. 28:1041-1059.

The Ordered Structure of the Crystalline Lens

J.R. Kuszak, B.A. Bertram, and J.L. Rae

The function of the crystalline lens is to focus light onto the retina. Consequently, the lens must be transparent. Lens transparency is in part made possible by the highly ordered arrangement of its uniform, terminally differentiated cells. It is an oversimplification to describe the lens as a biconvex spheroid comprised of crescent-like cells arranged end to end around a polar axis. To comprehend lens architecture it is necessary to consider embryonic lens development, the formation of primary and secondary fiber cells and the structure of lens sutures.

EMBRYONIC LENS DEVELOPMENT

Early in embryonic life, the surface ectoderm of the head directly overlying the optic vesicle is induced to thicken and become the lens placode which then begins to invaginate toward the forming optic cup until it pinches off completely as an inverted, hollow vesicle of epithelial cells.

Cells of the posterior half of the lens vesicle, the half facing the developing retina, are induced to terminally differentiate into primary lens fiber cells (Coulombre and Coulombre, 1963). These cells are referred to as "primary" fiber cells because they are the first such cells to be formed in the lens. They are called lens "fiber" cells, or simply fibers, because as a consequence of terminal differentiation the cuboidal posterior vesicular cells are transformed into very long, slender cells. The elongation of these cells is directed toward the anterior surface of the lens vesicle. Therefore, upon completion of primary fiber formation, the vesicular lumen is obliterated. At this time, the developing lens is a spherical mass of primary fibers covered by an epithelial monolayer on its anterior half.

LENS GROWTH OR SECONDARY FIBER CELL FORMATION

Lens growth commences with secondary fiber cell formation. Although secondary fibers are formed throughout life, the rate of lens growth is markedly reduced after birth. Lens growth occurs in the same manner in an adult lens as in an embryonic lens. Therefore, the pertinent changes in cellular shape and the establishment of order between secondary fiber cells can be demonstrated in either an embryonic or adult lens.

The simple epithelium overlying the anterior half of the lens actually has several properties of stratified epithelium. This epithelium can be separated into distinct regions or zones defined topographically and by the stage of the cell cycle that characterizes the majority of the cells in that zone (Harding et al., 1971). The central area of the lens epithelium is known as the central zone. Cells in this region of the epithelium are arrested in the G_1 phase of the cell cycle. They are considered to be maintained in the G_0 phase. These cells are very low cuboidal cells being much wider than they are tall (figure 1a). The lateral membranes of these cells are markedly infolded.

The pre-germinative zone is a narrow concentric region just beyond the central zone. Cells in this region are cuboidal epithelial cells being approximately as wide as they are tall (figure 1b). The lateral membranes of these cells, like the central zone cells, are markedly infolded. While the majority of the cells in this region are maintained in the G_0 phase, a very small number of these cells proceed through the cell cycle to mitosis.

The germinative zone is a narrow concentric zone beyond the pre-germinative zone. This zone of the lens epithelium has the characteristics of the basal germinal layer of a stratified

Figure 1. SEM microgrophs demonstrating the shape, size and surface morphology of cells from (a) the pre-germinative (PG) to central zone (C), (b) germinative (G) to pre-germinative zone, (c) transitional (T) to germinative zone and (d) bow region (BR) to transitional zone in the lens epithelium of an adult frog lens. Arrowheads delimit the expanded anterior ends of elongating fibers as they interface with cells in different regions of the lens epithelium. x1,000.

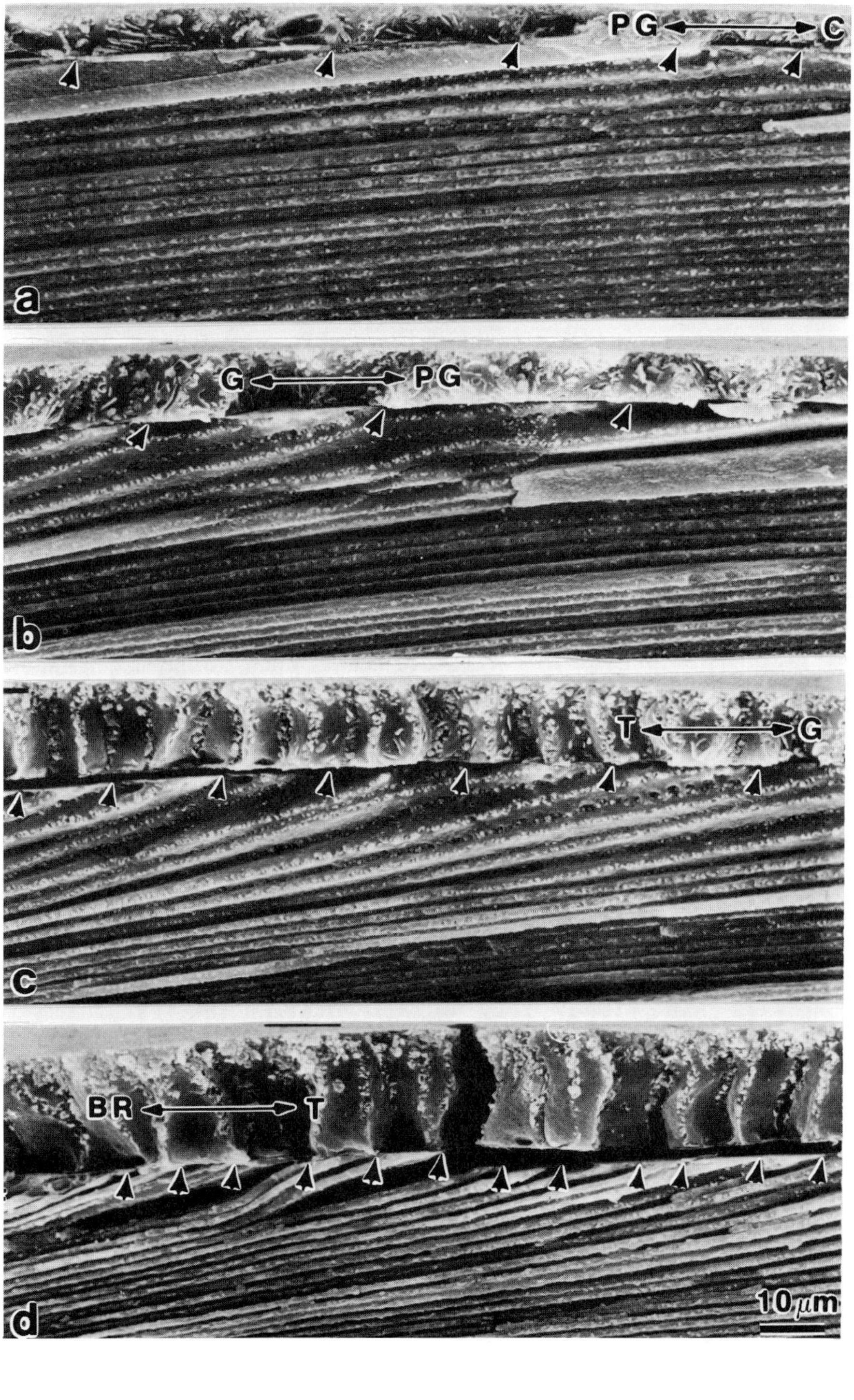
PG
C
a
G
PG
b
T
G
c
BR
T
d
10 μm

epithelium. Many of these cells undergo mitosis. Following mitosis, some of the daughter cells from this region continue through the cell cycle to terminally differentiate into secondary fibers. These fibers are then layered onto the existing primary fiber mass as strata of uniform crescent-like secondary fibers. As in the stratified epithelia of skin, the terminally differentiated cells (secondary fibers) eventually eliminate most of their internal organelles with the exception of the cytoskeleton. However, unlike skin or other stratified epithelia in which the outermost layer of terminally differentiated cells are eventually sloughed off, all the secondary fibers formed throughout life are permanently retained. Cells of the germinative zone are high cuboidal or low columnar epithelial cells (figure 1c). The lateral membranes of these cells unlike the central and pre-germinative cells, are not markedly infolded.

Continuous mitotic division in the germinative zone pushes the fiber precursor cells into the transitional zone, a concentric region just beyond the germinative zone (Messier and LeBlond, 1960; Mikulicich and Young, 1963; Hammar, 1965; and Beebe et al., 1980). Cells of this region are high columnar with uniform, hexagonal cross-sectional profiles. They are arranged in ordered rows (figure 1d). The basal surface is loosely attached to the specialized basement membrane of the lens epithelium known as the lens capsule. Of the six lateral faces of transitional zone cells, two are broad and four are narrow. The uniform shape and ordered alignment of fibers that characterizes all secondary fibers of the lens is established in this region.

When transitional zone cells are advanced to the lens bow, a region just above the equator, they are rotated 90° around their long axis (Rabl, 1900). During this rotation the transformation from high columnar epithelial cells with hexagonal cross-sectional profiles to thin fiber cells with hexagonal cross sectional profiles commences (figure 2a). As fiber elongation proceeds, the cells are referred to as young, forming, or elongating secondary fibers.

Figure 2. Photomontages made from SEM micrographs demonstrating (a) the elongation and attenuation of transitional zone cells as they are rotated 90° (arrows) in the bow region (BR) and (b) the expanded posterior ends (delimited by arrowheads) of elongating fibers interfacing with the lens capsule in an adult frog lens. x1,000.

BR
a
b
10 μm

As they elongate the young fibers become aligned so that they lie in a plane passed through the polar axis of the lens along their entire length. The broad faces of fibers are oriented parallel to the lens surface while the narrow faces are oriented at acute angles to the lens surface. The apical surface of an elongating fiber interfaces with the apical surfaces of overlying anterior epithelial cells while its basal surface remains in contact with the lens capsule (figure 2b). The broad and narrow faces of fibers have generally been referred to as long and short sides respectively (Duke-Elder and Wybar, 1963). However, since all the faces of any one fiber are equal in length , this terminology seems inappropriate. Instead, we propose that the different faces of fibers be referred to as broad and narrow, terms that more accurately distinguish the difference in size between these faces.

Secondary fibers elongate anteriorly and posteriorly at different rates (Kuszak and Rae, 1982). A secondary fiber reaches its anterior destination before reaching it posterior destination. An important consequence of the asymmetrical lengthening of fibers is that these cells are not equidistant from the lens surface along their length. As an example, in a lens measuring 3mm in diameter, the apical face of an elongating fiber interfaces with the apical faces of several central epithelial cells while its middle portion is located approximately 200 microns from the lens surface at the equator. Small molecules can move through gap junctions from epithelial cells to underlying elongating fibers and then pass 200 microns into the lens interior via simple diffusion along the relatively broad cross sectional area of the fiber cell (Rae and Kuszak, 1983). This route is more efficient than the serial passage through the gap junctions between the hundreds of fibers that comprise the peripheral 200 microns of a lens, or diffusion in the restricted extracellular clefts. This transport "short circuit" may be highly significant because the peripheral 200 microns of a lens with a 3mm diameter, represents about 30% of the total number of fibers in such a lens and these fibers are the most active metabolically.

Since fiber location is directly related to fiber age, the lens can be subdivided into regions according to cell age. An equatorial cross-section through a lens reveals radial rows or radial cell columns (RCCs; figure 3). Each RCC is composed of progressively younger fibers extending peripherally from the center of the lens to its surface. Cells of similar age constitute concentric shells or growth rings. The center of a lens is referred to as the lens nucleus, while the periphery of the lens is known as the lens cortex (figure 4). As a result of continuous lens growth, the lens cortex increases in volume with age (Vogt, 1921; Gallati, 1923). The nucleus of a lens can be further subdivided. The embryonic nucleus is comprised of all the original primary fiber cells. The fetal nucleus contains the embryonic nucleus and all the secondary fibers formed prior to birth. The adult nucleus is comprised of the fetal and embryonic nuclei as well as all the secondary fibers formed prior to sexual maturation.

FIBER DIMENSIONS

Fiber thickness is the breadth of a fiber between its two broad faces. Fiber width is the distance between the apices formed by opposite pairs of narrow faces (figure 3). While fiber thickness varies along a fibers' length, it is relatively constant as a function of radial depth (Kuszak and Rae, 1982). The circumference of a fiber is related to its thickness, width and length. Consequently, the reduction in a fiber's circumference toward its ends, known as fiber taper, is primarily due to a decrease in fiber width. The amount of fiber taper is related to lens sutures and will be discussed later. Variation in fiber circumference along fiber length is in the order: middle region > posterior end > anterior end. This in part explains the variation in the degree of flattening or less anterior curvature in lenses of different species (Duke-Elder and Wybar, 1963).

In comparison, the variation in elongating fiber circumference is in the order: anterior end > posterior end > middle region. Young fibers flare at their ends as they elongate. The ends of fibers taper after completing the elongation process. The anterior expansions of young fibers vary in size in accordance with the region of the epithelium that they contact. The size of the contact is in the order: pre-germinative zone > germinative zone > central zone > transitional zone (figure 1). Since the area of an anterior expansion of a young fiber is always greater than the area of the apical surface of an overlying epithelial cell, several epithelial cells interface with a single anterior expansion. The epithelial cells and the elongating fibers are electrotonically coupled, presumably through gap junctions (Rae and Kuszak, 1983). Consequently, these gap junctions must be formed and reformed as the young fibers contact cells in different regions of the anterior epithelium during their elongation toward the anterior suture. The posterior expansions of elongating fibers are largest in the initial stages of elongation and then become progressively smaller as they approach the posterior suture.

Fiber length varies as a function of fiber depth or age. Fiber length can be approximated as πr, where r is equal to the radius at the depth where the fiber is located. Therefore, the increment in fiber length between fibers in successive growth rings is $\pi(r + k)/\pi r$, where k is equal to the thickness of the fiber in the next youngest growth ring. As described earlier, fiber thickness and thus k is relatively constant at all radial depths. Fiber length is variable within any growth ring. This variation, like fiber cell taper, is related to lens sutures and will be discussed later.

As a lens grows, its size increases radially. An equatorial cross section through a lens reveals that the increase in lens size is not accomplished by the addition of proportionately larger fibers (figure 3). Rather, it results from the addition of more cells in each successive growth ring (figure 5).

FIBER SURFACE MORPHOLOGY

Epithelial cells are characteristically held or bound together by adhering junctions (zonula occludens or tight junctions, zonula adherens, macula adherens or desmosomes and fascia adherens) between their lateral membranes. Although the broad and narrow faces of fibers are analogous to the lateral membranes of the cuboidal epithelial cell from which they developed, there is a complete absence of adhering junctions between fibers. Instead, lens fibers are held together by a system of highly developed lateral interdigitations. Interdigitations of lateral membrane in

Figure 3. SEM micrographs of cortical fibers in cross section produced by dry fracturing a critically point dried adult frog lens along its equator. A representative growth ring (GR), radial cell column (RCC), narrow faces (NF), broad faces (BF), fiber thickness (FT) and fiber width (FW) are indicated. x4,000.

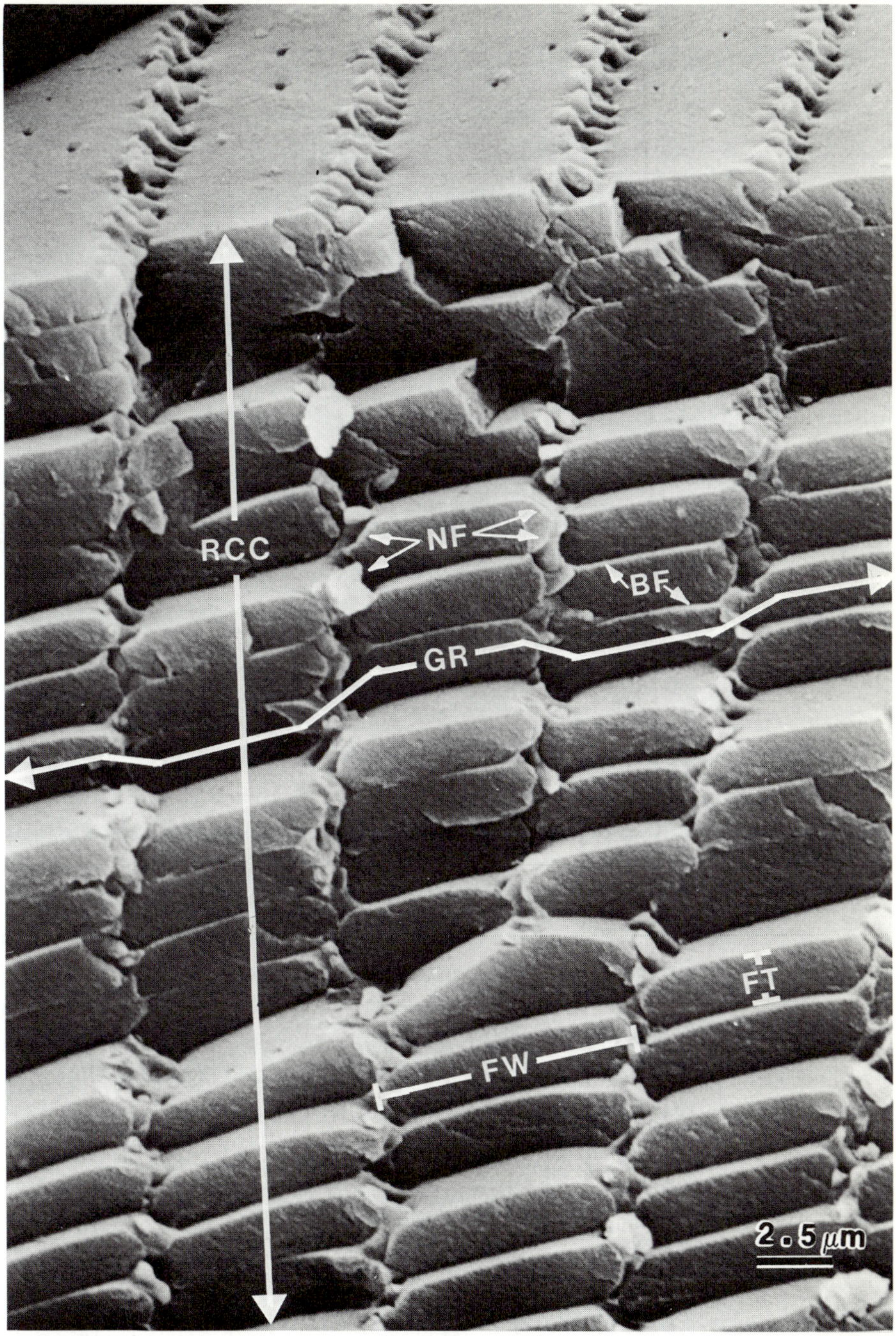
RCC
NF
BF
GR
FT
FW
2.5 μm

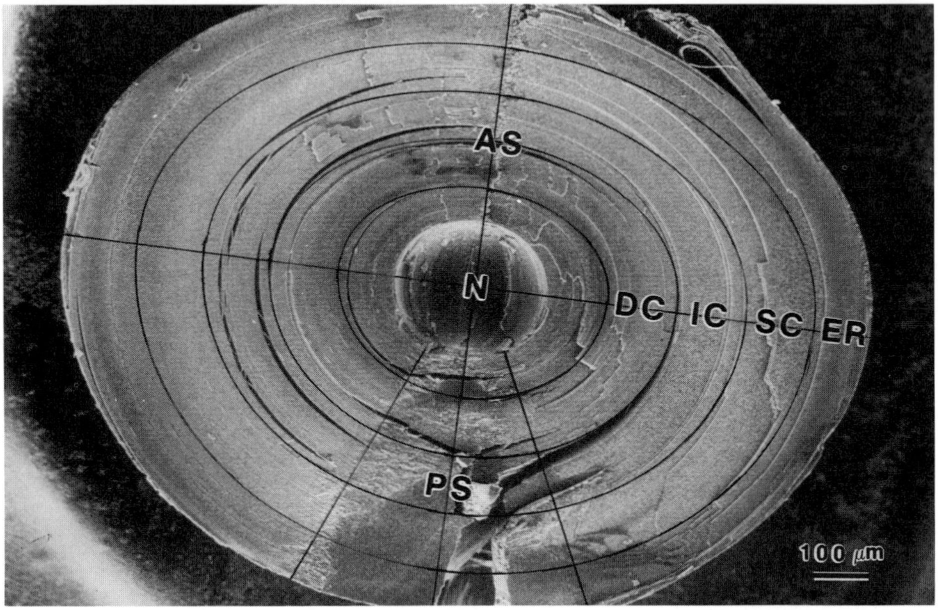

Figure 4. Low magnification SEM micrograph of one half of a critically point dried adult frog lens produced by dry fracturing the lens along its polar axis. The concentric lamellar arrangement of fibers from suture to suture results in different aged regions known as the nucleus (N), deep cortex (DC), intermediate cortex (IC), superficial cortex (SC) and elongating region (ER). Anterior suture plane (AS) and posterior suture plane (PS). x41.

epithelial cells generally provide weak adhesion between neighboring cells and/or increase in membrane surface area. The lateral interdigitations of lens fibers are more numerous, more uniform and more regularly arrayed than the lateral interdigitations of any other epithelial cells. The location of lateral interdigitations and their shapes are basically the same in all lenses. The shapes and topographical locations of lateral interdigitations between fibers are best demonstrated by scanning electron microscopy (Kuwabara, 1970 and 1975; Hansson, 1970; Dickson and Crock, 1972 and 1975; Farnsworth et al., 1974; Harding et al., 1976; Hollenberg et al., 1976; Nelson and Rafferty, 1976; Kuszak et al., 1980, 1983; Willekens and Vrensen, 1981; Hoyer, 1982). At the angles formed by two narrow faces lateral interdigitations are found in the shape of a ball atop a short stalk. They fit into complementary shaped sockets formed between fibers of adjacent growth rings (figure 6). Flap-like evaginations are found at the angles formed by narrow and broad faces. They fit into complementary shaped shallow imprints

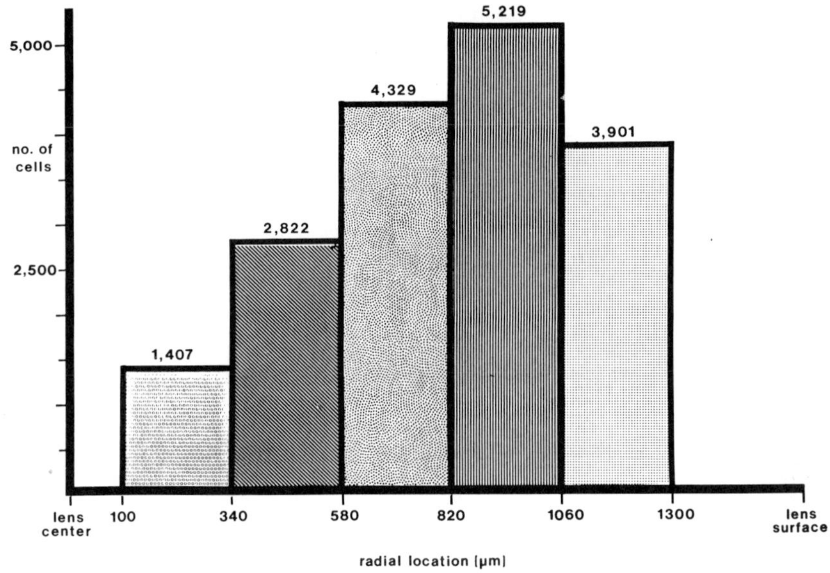

Figure 5. Histogram plotting the average number of fibers contained in approximately 30 longitudinal degrees from the center of an adult frog lens to 0° latitude (n = 8).

formed on the narrow faces of fibers in adjacent growth rings (figure 7). Both balls and flaps and their sockets and imprints respectively, are repeated in an alternating pattern along the length of fibers. A review of the literature shows that balls and flaps of fibers have frequently been considered identical. This is not the case. A ball is always fit into a socket or a cavity whereas a flaps is always fit into a shallow imprint. The problem of misidentification between balls and flaps seems to arise by viewing these interdigitations in two-dimensionsal micrographs. When these interdigitations are viewed in stereomicrographs it is clear that they are structurally different (see figure 6 and 7). Balls and sockets also are found as evaginations of broad face membranes. However, these balls and sockets are considerably smaller than those found at the angles of narrow faces (figure 8). Lateral interdigitations are found along the entire length of mature secondary fibers. They do not occur along the entire length of elongating fibers. Rather, they are only found at the anterior and posterior flared ends of elongating fibers.

The lateral interdigitations between fibers are generally

referred to as interlocking devices. This terminology seems appropriate since they are the only adhering type "junction" between these epithelial cells. It is not likely that the sole function of interlocking devices is to maintain fiber aligmnent when lens deformation occurs during accomodation (Barbaraschew, 1892; Hosch, 1901; Farnsworth et al., 1974; Willekens and Vrensen, 1981; Hoyer, 1982). Perhaps the most complex interlocking devices are found in lenses that do not change shape during accomodation such as fish lenses (figure 8).

Both the broad and narrow faces of fibers are typically smooth and much of the surface is specialized as gap junctions (Bloemendal, 1972; Benedetti et al., 1976; Kuszak et al., 1978; Goodenough, 1979). In some lenses (frog, monkey and human) both the broad and narrow faces of nuclear fibers are undulate (Dickson and Crock, 1972 and 1975; Kuwabara, 1975; Sakuragawa et al., 1975). The undulations of this membrane have been considered to be another type of interlocking device referred to as tongue and grooves (Dickson and Crock, 1972 and 1975). The undulating membrane was generally considered to be an age related change because it was only seen in nuclear fibers (Kuwabara, 1975; Kuszak and Rae, 1982). It is known that the broad faces of cortical fibers of human neonatal and adult guinea pig lenses also are undulate (Kuszak et al., 1983; figure 9). Recently, it has been shown that the undulate membrane in human lenses is related to the square array junction (Lo and Harking, 1984; Costello et al., 1984). It has been proposed that the square array junction may represent a breakdown or alteration in the original gap junctions between fiber cells resulting in an alteration of membrane from smooth to undulating. Undulate membrane is inconsistent with gap junctional structure (Goodenough, 1978).

Figure 6. SEM stereomicrographs ($\pm 6^{o}$) demonstrating the ball-like interlocking devices (arrows) and complementary sockets (arrowheads) of fibers in an adult rat lens. x3,800.

Figure 7. SEM stereomicrographs ($\pm 6^{o}$) demonstrating the flap-like interlocking devices (arrows) and complementary imprints (arrowheads) of fibers in an adult rat lens. x3,000.

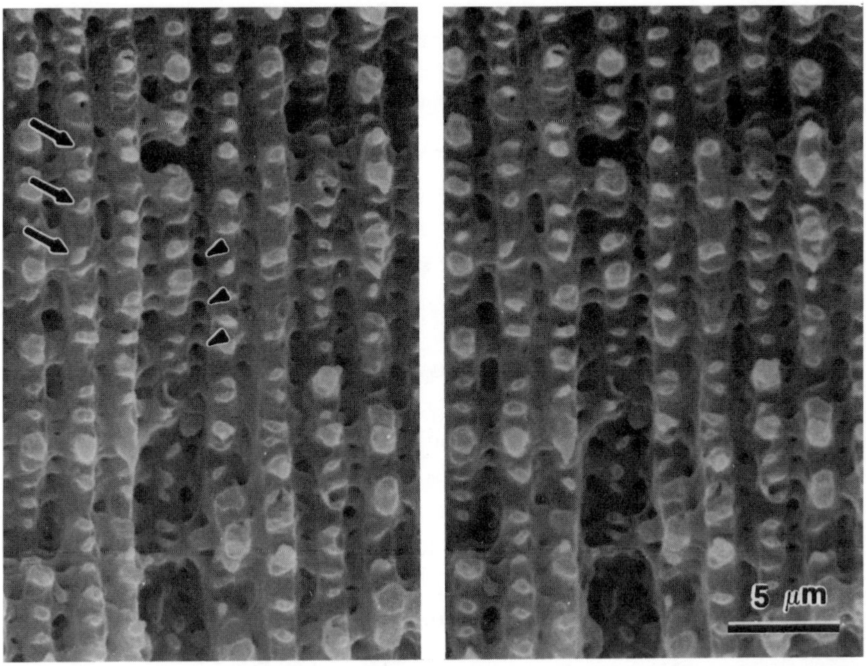
5 μm

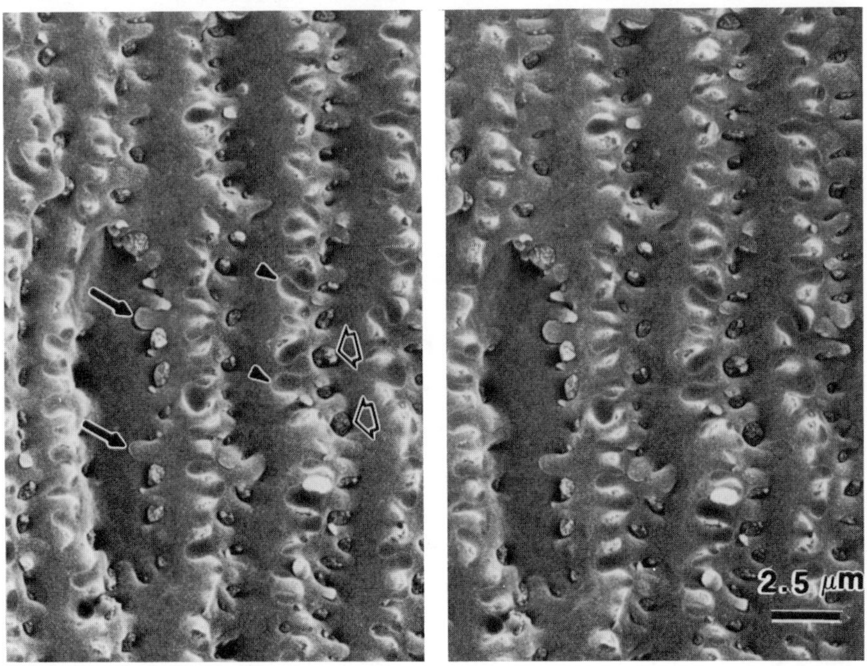
2.5 μm

LENS SUTURES

The interlocked ends of all secondary fiber cells form the lens sutures. Different types of sutures are formed in different species (Rochon-Duvigneaud, 1943; Duke-Elder and Abrams, 1970; de Jong, 1981; Kuszak et al., 1984). The determination of suture type is made during a specific period in embryonic life (Woerdemaan, 1934). The four types of lens sutures are in order of increasing complexity: the umbilical, line, y and star suture. Since lenses are spheroids, an understanding of suture formation and type is facilitated by using terminology similar to that used to describe geometric parameters on terrestrial globes (Kuszak et al., 1984). Latitudinal angles and longitudinal meridians are particularly appropriate for describing lens suture formation and type. The lens can be considered as a coordinate system in which the longitudinal angles are generated by fiber cells (halves of a great circle of a sphere) which rotate a full 360° clockwise around an axis defined by the anterior and posterior poles of the lens. Zero degrees is defined when the plane of the circle is oriented vertical to the visual axis. It is these longitudinal angles that are used to describe the orientation of suture branches. Latitudinal angles are defined as 0° at the equator, 90° at the anterior pole and -90° at the posterior pole. It is these latitudinal angles that are used to describe the position or termination points of fiber cell ends at the suture branches. Obviously, any point on the lens surface or in a growth ring can be described by the intersection of a latitude and a longitude. When referring to a point on the lens by convention the latitudinal location is referred to first and the longitudinal location second. Viewed in the anatomical position, the circumference of all lenses is a circle defined by the lens equator. When viewed along the equatorial plane, the circumference of lenses is generally an asymmetric ellipse with the anterior half having a greater radius of curvature than the posterior half.

Figure 8. SEM stereomicrographs ($\pm 6^{o}$) demonstrating the elaborate and highly regular interlocking devices of fibers in an adult fish (rainbow trout) lens. x3,000.

Figure 9. SEM micrograph demonstrating the undulate membrane of fibers in an adult guinea pig lens. x3,000.

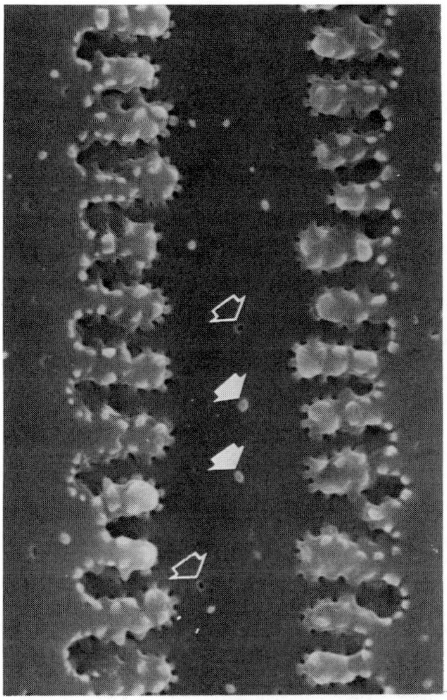

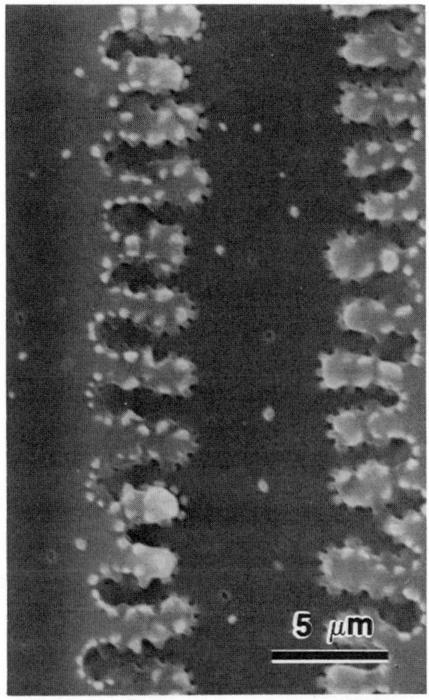
5 μm

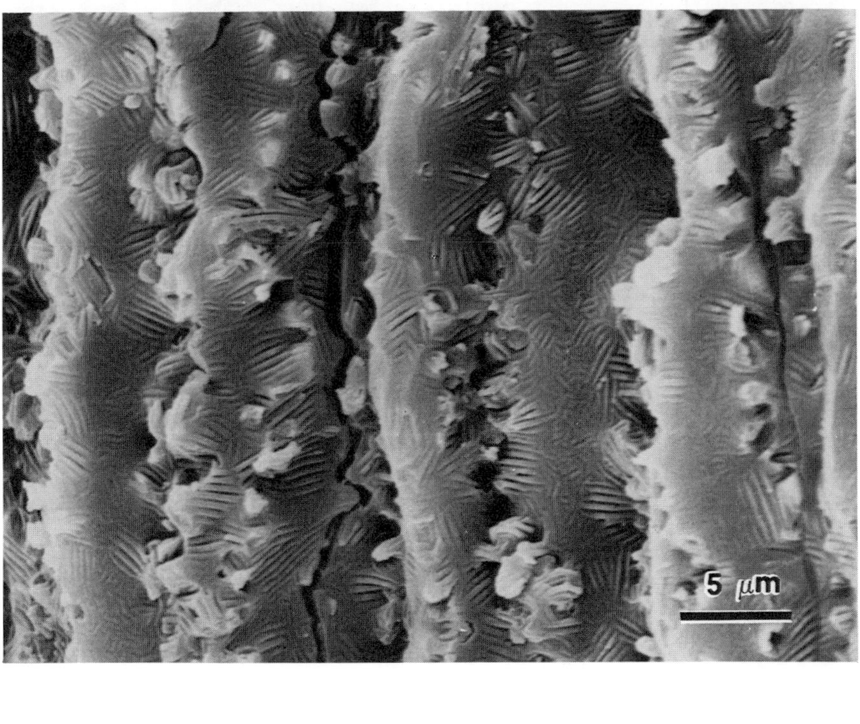
5 μm

UMBILICAL SUTURES

The simplest suture type is an umbilical or central point suture. This suture is characteristic of avian, reptilian, some amphibian (salamander), and certain teleost lenses. In lenses with umbilical sutures, fiber cells are essentially meridians; that is each fiber cell is half of the circumference of its respective growth ring. These fibers are oriented along their entire length in a plane that passes through the polar axis of the lens and the length of all fibers in any growth ring is identical.

LINE, Y AND STAR SUTURES

Line sutures are characteristic of many teleosts, sharks and rays, certain salamanders, amphibians (frog) and some small mammals (mice, rats, guinea pigs and rabbits). Y sutures are also seen in younger lenses of large mammals (kitten, puppy, monkey and human neonate). The most complex suture type, the star suture, is seen in large adult mammals (cat, dog, monkey and human). In lenses with line, y or star sutures, the fiber cells do not extend from pole to pole and are therefore not meridians. Rather, fiber cells in these lenses interlock along anterior and posterior latitudinal arcs or suture branches that extend from the lens poles toward the lens equator. The end of a suture branch nearest a lens pole is the proximal end, while the end of a suture branch nearest the lens equator is the distal end. In lenses with line sutures, the two anterior suture branches extend respectively from the anterior pole to 60°, 0° and 60°, 180°. The posterior suture branches extend respectively from the posterior pole to -60°, 90° and -60°, 270° (figure 10). Hence, the name line suture because in the anatomical position, the anterior and posterior suture branches, appear respectively as vertical and horizontal lines on the anterior and posterior surfaces. Similarly, y sutures are so named because in

Figure 10. Photographs of a 50:1 scale model of a lens with line sutures taken at an angle to its polar axis to demonstrate the variable fiber cell end curvature in a single growth ring. The model was based on stereomicrographs of a 3mm diameter adult frog lens. Each fiber is representative of 20 fibers in the actual lens. (a) Model with fibers contained in 90 longitudinal degrees and (b) in a complete growth ring located approximately 3mm from the lens center.

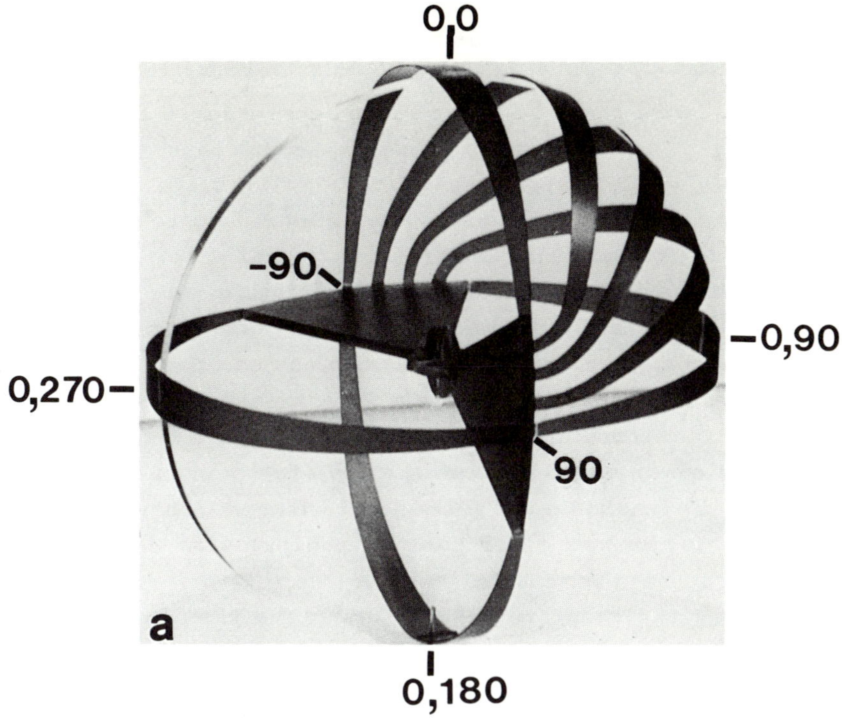
0,0
-90
0,270
-0,90
90
a
0,180

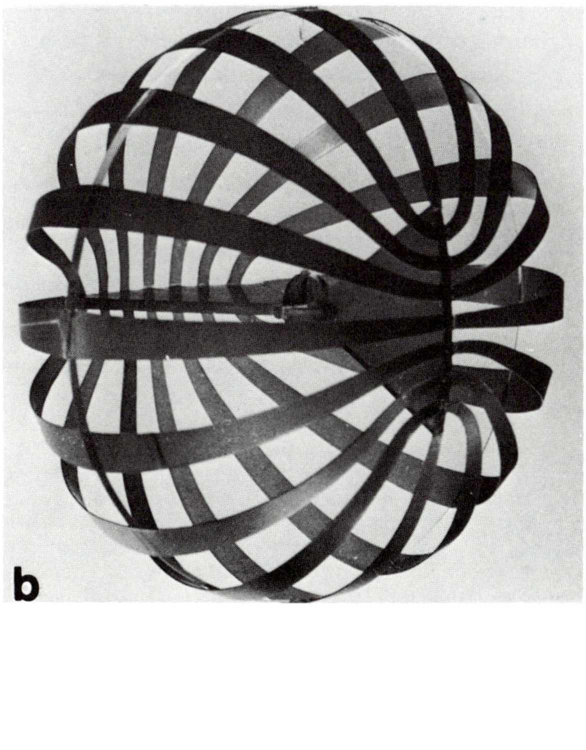
b

the anatomical position there are three anterior suture branches that form a y on the anterior lens surface and three posterior branches that form an inverted y on the posterior lens surface (figure 11). Note that the anterior and posterior suture branches of line and y sutures are rotated to each other by 90 and 60 longitudinal degrees, respectively. Star sutures are characterized by as many as nine branches and seven or more sub-branches on the anterior and posterior surfaces (Kuszak et al., 1984). The distal ends of branches and sub-branches of star sutures are much farther from the poles than the branches of line and y sutures. The anterior and posterior branches and sub-branches of star sutures are rotated to each other approximately 20 longitudinal degrees. The exact amount of rotation is difficult to determine because the branches and sub-branches in star sutures are not confluent.

In lenses with line, y or star sutures, there are three distinct variations in fiber cell morphology when compared with fibers from lenses with umbilical sutures. First, fibers are not aligned along their entire length in a plane passing through the polar axis of the lens. Second, all fibers in any growth ring are not in general of equal length. Third, fiber taper varies in proportion to the number of suture branches.

Fiber cell curvature refers to the concave curvature along the length of a fiber. The ends of fibers in lenses with line, y or star sutures turn away from the lens polar axis in order to interlock at different points along a suture branch. This curvature of fiber ends away from the lens polar axis can be referred to as fiber cell end curvature. Because the anterior and posterior sutures are rotated with respect to each other, fiber cell end curvature is in opposite directions anteriorly and posteriorly (see figures 10, 11 and 12).

In lenses with line, y or star sutures all fiber cells in any growth ring are not of equal length because they interlock at different points along suture branches. Furthermore, since fiber

Figure 11. Photographs of a 100:1 scale model of a lens with y sutures taken at an angle to its polar axis to demonstrate the variable fiber cell end curvature in a single growth ring. The model was based on stereomicrographs of a 3mm diameter adult rat lens. Each fiber is representative of 10 fibers in the actual lens. (a) Model with fibers contained in 60 longitudinal degrees and (b) in a complete growth ring located approximately 3mm from the lens center.

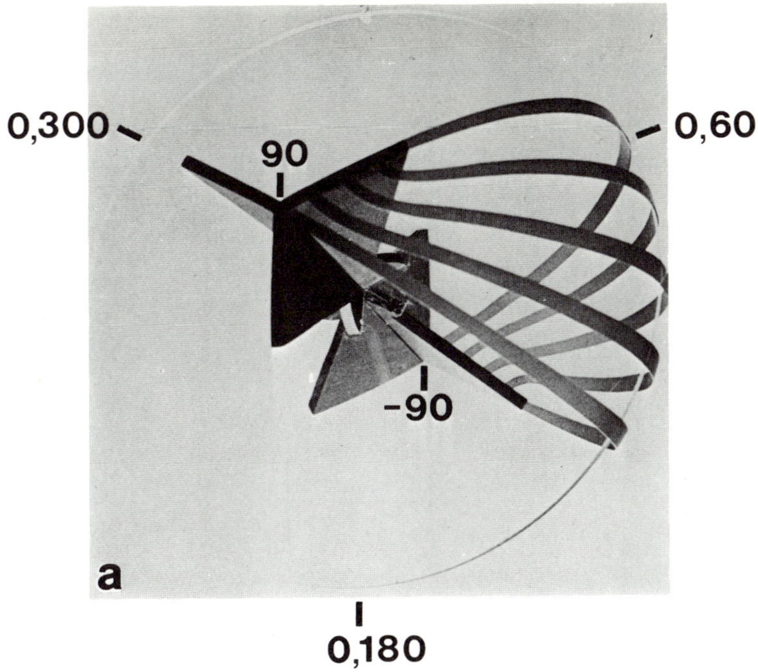
0,300
0,60
90
-90
a
0,180

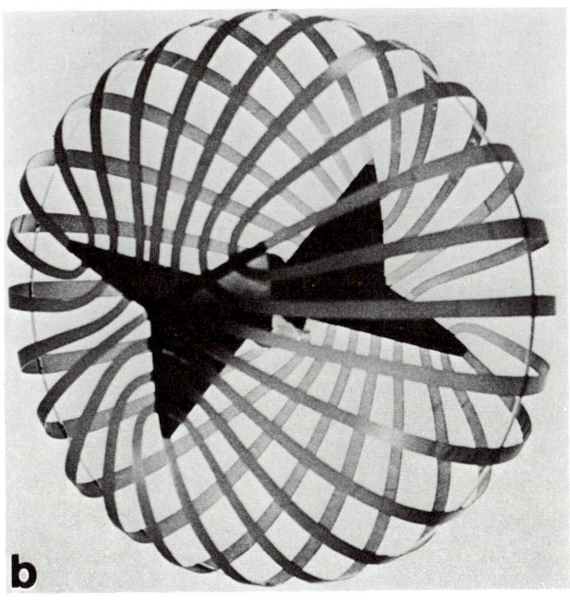
b

cell end curvature increases fiber length, the length of fibers in lenses of comparable size but with different sutures will not be equal. As the number of suture branches increases the longitudinal distance between anterior and posterior suture branches decreases. Therefore, while the range of fiber length in comparable growth rings from lenses of identical size will be the same, fiber length in any growth ring produces an approximate sine wave curve that is a function of the longitudinal location of a fiber cell (figure 13). The number of times the sine wave curve is repeated in a period (360 longitudinal degrees or a complete growth ring) is identical to the number of suture branches.

Fiber cell taper, like fiber cell length and fiber cell end curvature is also related to the number of suture branches. In lenses with umbilical sutures, fibers are fusiform because all fibers interlock at the poles. In lenses with line and y sutures, fibers need not taper as severely because they interlock along arc lengths that measure approximately thirty latitudinal degrees. In lenses with line sutures, the reduction in fiber width (the ratio of fiber width at the equator to fiber width at the sutures) is 3:1. In lenses with y sutures, the reduction in fiber width from equator to sutures is 2:1. In lenses with star shaped sutures, fibers flare rather than taper. The flaring of fiber ends is imperative because these lenses fibers are contained in approximately 20 longitudinal degrees at the equator but interlock at suture branches that are 45 latitudinal degrees or more in length.

The distal ends of fibers in all lenses, irrespective of suture type, are slightly enlarged in both thickness and width as they interlock to form sutures (figure 14). Sutures are formed by the overlap of the enlarged distal ends of fibers rather than the simple fiber abutment end to end. This overlapping results in three-dimensional suture planes extending from the primary fiber cell mass to the periphery of the adult lens.

Figure 12. Photographs of the scale models of lenses with (a) line sutures and (b) with y sutures taken in the anatomical position (along the polar axis).

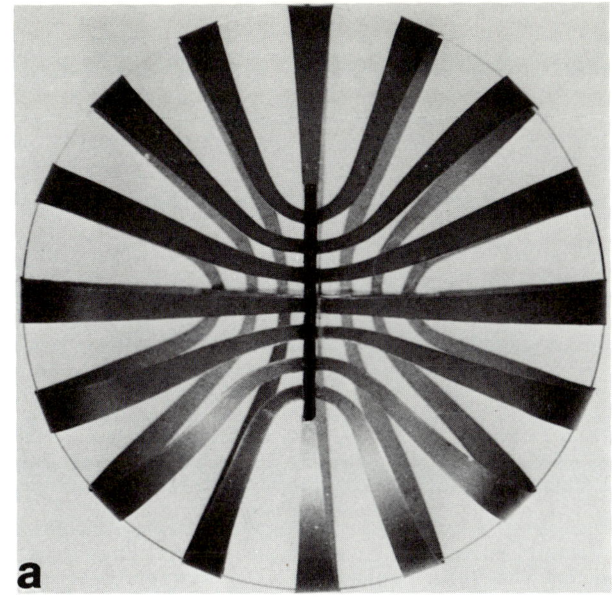

a

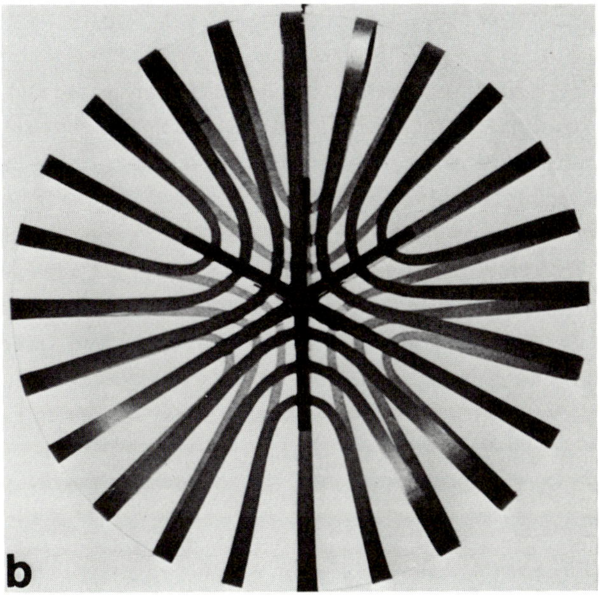

b

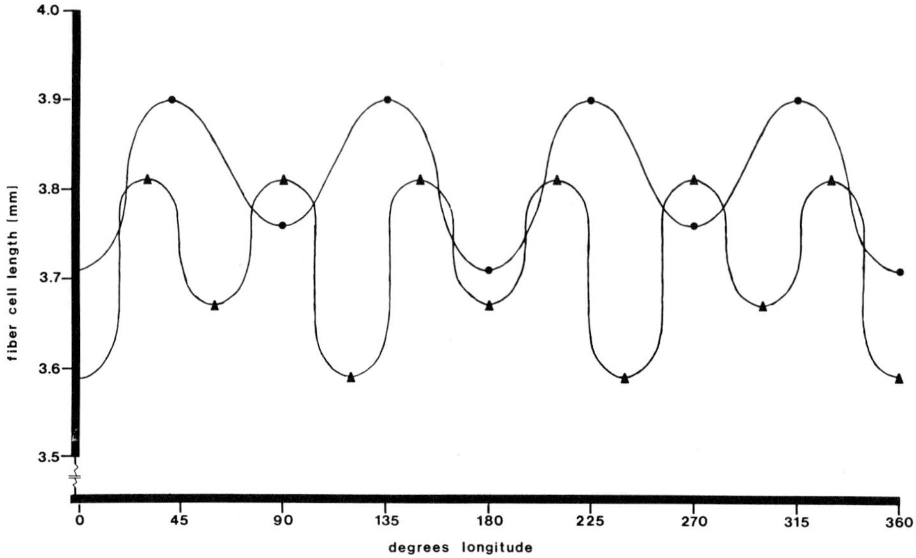

Figure 13. Graph plotting the length of fibers in a growth ring located 3mm from the lens center in a lens with line sutures ( ) and in lens with y sutures ( ) against the equatorial (longitudinal) location of the fibers.

SUMMARY

Throughout life, cells from the germinative zone of the lens epithelium are induced to terminally differentiate and become lens fibers. As a consequence of terminal differentiation some cells from the germinative zone are transformed from cuboidal epithelial cells into long, slender fibers.

The lens grows by the addition of these fibers onto its periphery. The circumference of young fibers varies along their length. They have asymmetrically flared ends, the anterior ends always being larger than the posterior and mid portions. Furthermore, the extent of flaring of an anterior end of a young fiber varies as it elongates toward the anterior suture to interface with the apical surfaces of epithelial cells in different regions of the anterior epithelium. Thus gap junctions which are presumed from

Figure 14. (a) SEM stereomicrographs ($\pm 6^{o}$) of the interlocked anterior ends of fibers at a portion of an anterior suture in an adult frog lens. x907. (b) SEM micrograph of the anterior apical ends of fibers (between brackets) at a portion of an anterior suture. x3,000.

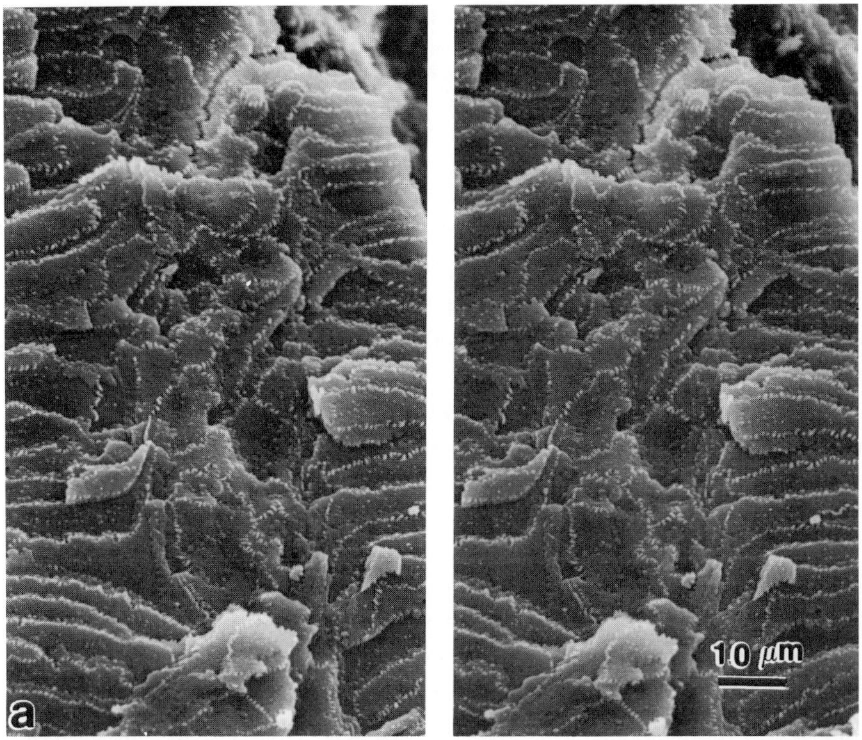
a
10 μm

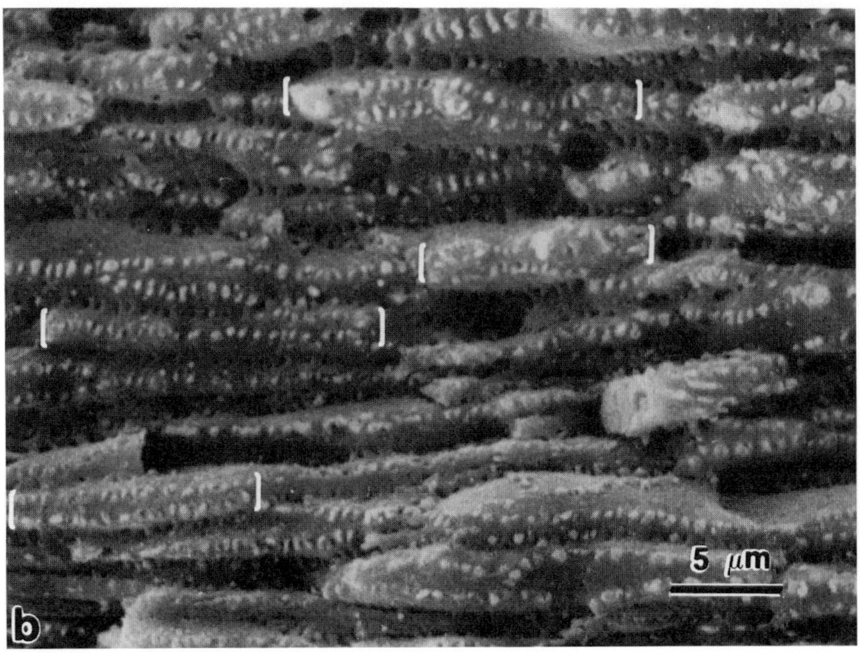
b
5 μm

electrophysiological experiments to conjoin these cells must constantly be formed and reformed. The anterior expansions are largest as young fibers interface with cells in the pre-germinative zone.

Upon completion of elongation, fiber shape is altered. The variation in fiber circumference along length is in the order: middle portion > posterior end > anterior end. The alteration in fiber shape after elongation in part explains why embryonic lenes are spherical and adult lenses are spheroidal. The asymmetric variation in fiber circumference along fiber length in part explains why adult lenes are generally asymmetric spheroids.

Fibers are arranged in highly ordered radial cell columns extending from the lens nucleus to the lens periphery. All the fibers formed at approximately the same time make up growth rings. The fibers are held together by uniformly shaped and regularly arrayed interdigitations between adjacent broad and narrow faces. These interdigitations are commonly referred to as interlocking devices.

The ends of all secondary fibers interlock to form sutures. Fibers either interlock at the lens poles (umbilical suture) or taper and fiber cell end curvature are all related to the number of suture branches.

ACKNOWLEDGEMENTS

This work supported by National Institutes of Health grant nos. EY 04935 to JRK and EY 03282 to JLR, the Louise C. Norton Trust and the Regenstein Foundation. The authors wish to thank Mr. James Brainer, Mr. K. C. Carney, Ms. Linda Jones, Ms. Suzanne Imherr-McManiss, Marian S. Macsai-Kaplan M.D. and Mr. Y. H. Shek for their assistance in the studies on which this work is based.

REFERENCES

Barabaschew, P. 1892. Bietrag zur anatomie der lense. Albrecht von Graefe's Archives zur Ophthalmologie 38: 1-14.

Beebe, D. C., M. C. Johnson, D. E. Feagans, and P. J. Compart. 1980. The mechanism of cell elongation during lens fiber cell differentiation. In "Ocular Size and Shape Regulation During Development." (Hilfer, S. R. and J. B. Sheffield., eds.) pp 79-98.

Benedetti, E. L., I. Dunia, C. J. Bentzel, A. J. M. Vermorken, M. Kibbelaar and H. Bloemendal. 1976. A portrait of plasma membrane specializations in lens epithelium and fibers. Biochim. et Biophys. Acta 457: 353-384.

Bloemendal, H., A. Zweers, F. Vermorken, I. Dunia and E. L. Benedetti. 1972. The plasma membranes of eye lens fibers. Biochemical and structural characterization. Cell Differ. 1: 91-106.

Costello, M. J., T. J. McIntosh, and J. D. Robertson. 1984. Square-array fiber cell membranes in the mammalian lens. In "Proceedings of the 42nd Annual Meeting of the Electron Microscopy Society of America." (Bailey, G. W., ed.) pp 126-129.

Coulombre, J. and A. J. Coulombre. 1963. Lens development: Fiber elongation and lens orientation. Science 142: 1489-1490.

de Jong, W. W. 1981. Evolution of lens and crystallines. In "Biology of the Eye Lens." (Bloemendal, H., ed) J. Wiley and Sons Inc., New York.

Dickson, D. H. and G. W. Crock. 1972. Interlocking patterns of primate lens fibers. Inves. Ophthalmol. 13: 809-815.

Dickson, D. H. and G. W. Crock. 1975. Fine structure of primate lens fibers. In "Cataract and Abnormalities of the Lens." (Bellows, J. G., ed.) pp 49-58. Grune and Stratton, New York.

Duke-Elder, S. and K. C. Wybar. 1961. The refractive media: the lens. In "System of Ophthalmology." vol. II. (Duke-Elder, S., ed.) pp 309-324. The C. V. Mosby Co., St. Louis.

Duke-Elder, S. and D. Abrams. 1970. The refraction of light. In "Systems of Ophthalmology." vol. V. (Duke-Elder, S., ed.) pp 74-78. C. V. Mosby Co. St. Louis.

Farnsworth, P. N., S. C. J. Fu, P. G. Burke and I Bahaia. 1974. Ultrastructure of rat lens fibers. Invest. Ophthalmol. 13: 274-279.

Gallati, J. 1923. Die relativen Dickenwerte von Rinde und Kern der menschlichen Linse in verschiedenen Lebensaltern. Atschr. f. Augenh. (Berlin) 51: 133-144.

Goodenough, D. A., D. L. Paul and K. E. Culbert. 1978. Correlative gap junction ultrastructure. Birth Defects. 14(2): 83-97.

Goodenough, D. A. 1979. Lens gap junctions: A structural hypothesis for nonregulated low-resistance intercellular pathways. Invest. Ophthalmol. Vis. Sci. 11: 1104-1122.

Hammar, H. 1965. An autoradiographic study on cell migration in the eye lens epithelium from normal and alloxan diabetic rats. Acta. Ophthal. (Kobenhavn) 14(2): 83-97.

Hanna, C. and J. E. O'Brien. 1961. Cell production and migration in the epithelial layer of the lens. Arch. Ophthal. (Chicago) 66: 103-107.

Hansson, H. A. 1970. Scanning electron microscopy of lens of the adult rat. Z. Zellforsch. 107: 187-198.

Harding, C. V., J. R. Reddan, N. J. Unakar and M. Bagchi. 1971. The control of cell division in the ocular lens. Int. Rev. Cytol. 31: 215-300.

Hollenberg, M. J., J. P. H. Wyse, and B. J. Lewis. 1976. Surface morphology of lens fiber from eyes of normal and microphthalmic (Browman) rats. Cell Tissue Res. 167: 425-428.

Hosch, G. 1901. Ein Fall von sog. cortikaler Hemianopsie und Alexie. Ztschr. f. Augenh. (Berlin) 5: 5-14.

Hoyer, H. E. 1982. Scanning electron-microscopic study of lens fibers of the pig. Cell Tissue Res. 224(1): 225-232.

Kuszak, J., H. Maisel and C. V. Harding. 1978. Gap junctions of chick lens fiber cells. Exp. Eye Res. 27: 495-498.

Kuszak, J., J. Alcala, and H. Maisel. 1980. The surface morphology of embryonic and adult chick lens-fiber cells. Am. J. Anat. 159: 95-410.

Kuszak, J. R. and J. L. Rae. 1982. Scanning electron microscopy of the frog lens. Exp. Eye Res. 35: 499-515.
Kuszak, J. R., M. S. Macsai and J. L. Rae. 1983. Stereo scanning electron microscopy of the crystalline lens. Scan. Elec. Micr. 111: 1415-1426.
Kuszak, J. R., B. A. Bertram, M. S. Macsai and J. L. Rae. 1984. Sutures of the crystalline lens: A review. Scan. Elec. Micr. 111: 1369-1378.
Kuwabara, T. 1970. Surface structure of the eye tissue. Proc. III Annual Scanning Electron Microscope Symposium. pp 185-192. Chicago.
Kuwabara, T. 1975. The maturation of the lens cell: A morphological study. Exp. Eye Res. 20: 427-443.
Langman, J. 1959a. The first appearance of specific antigens during the induction of the lens. J. Embryol. and Exper. Morphol. 7: 193-202.
Langman, J. 1959b. Appearance of antigens during development of the lens. J. Embryol. and Exper. Morphol. 7: 264-274.
McKeehan, M. S. 1951. Cytological aspects of embryonic lens induction in the chick. J. Exper. Zool. 117: 31-64.
Messier, B. and C. P. LeBlond. 1960. Cell proliferation and migration as revealed by radioautoradiography after injection of thymidine-^{3}H into male rats and mice. Amer. J. Anat. 106: 247-285.
Mikulicich, A. and R. W. Young. 1963. Cell proliferation and displacement in the lens epithelium of young rats injected with tritiated thymidine. Invest. Ophthal. 2: 344-354.
Rabl, C. 1900. Concerning the structure and development of the lens. Atschr. F. Wissenensch. Zool. 66: 1-138.
Rochon-Duvigneaud, A. 1943. "The Eyes and Vision of Vertebrates." Masson and Cie., Paris.
Sakuragawa, M., T. Kuwabara, J. Kinoshita, and H.N. Fukui. 1975. Swelling of the lens fibers. Exp. Eye Res. 21: 381-394.
Smelser, G. K. 1965. Embryology and morphology of the lens. In: Symposium on the lens. J. E. Harris (Ed.), The C. V. Mosby Co., St. Louis, Mo., 22-34.
van Doorenmaalen, W. J. 1981. "The developmental mechanics of the lens." (Bloemendal, W., ed.) pp 415-435.
Vogt, A. 1921. Interstitial spaces of lens under certain forms of illumination. Schweiz. Med. Wehnschr. 51: 265-268.
Willenkens, B. and G. Vrensen. 1981. The three-dimensional organization of lens fibers in the rabbit. A scanning electron microscopic reinvestigation. Albrecht von Graefe's Arch. Klin. Exp. Ophthal. 216(4): 275-289.
Woerderman, M. W. 1962. Eye lens development and some of its problems. Proc. Koninklijke Nederlandse Akademie van Wetenschappen, series C, 65: 145.

The Main Intrinsic Polypeptide and Intercellular Communication in the Ocular Lens

Paul G. FitzGerald

We discuss here, the role of the Main Intrinsic Polypeptide (MIP) of the ocular lens fiber cell membrane, and its putative role in intercellular communication in the lens. In order to best understand the evidence which has been used in the consideration of MIP function, brief and non-comprehensive reviews of both the gap junction and lens biology are provided.

The gap junction (nexus, macula communicans, communicating junction) is almost universal in its distribution. It has been found coupling cells in virtually all tissues examined, throughout the animal kingdom. Notable exceptions are skeletal muscle cells, and blood cells (Dewey and Barr, 1964; Friend and Gilula, 1972; Gilula, et. al., 1972; Bennett and Goodenough 1978).

A vast literature of circumstantial evidence exists which implicates the gap junction as the morphological substrate which mediates processes such as ionic coupling and metabolic cooperation between adjacent cells (Dewey and Barr 1962; Bennett, et. al., 1963; Lowenstein and Kanno, 1964; Barr, et. al., 1965; Payton, et. al., 1969; Bennett and Goodenough, 1978; Lowenstein, 1979)) Our understanding of how the gap junction mediates these processes stems largely from ultrastructural and physiologic studies (cf. review Bennett and Goodenough, 1978) which suggest that the junction is a collection of intercellular channels which passively permit the diffusion of molecules under approximately 900 daltons between coupled cells (Simpson, et. al., 1977; Flagg-Newton, et. al., 1979; Schwarzmann, et. al., 1981). Notably this includes most metabolic intermediates and substrates, such as simple sugars, amino acids, nucleotides, as well as ions responsible for the existence of transmembrane electrical potentials.

The physiologic evidence which points towards a communication role for the gap junction has been gathered from a wide variety of tissues, both in situ and in vitro, using direct electrical recording with microelectrodes, iontophoresis of fluorescent dyes (Bennett, et. al., 1963;Lowenstein and Kanno, 1964; Asada and Bennett, 1971; Rose and Lowenstein, 1975; Turin and Warner, 1977, and 1980; Rose and Rick, 1978), and the intercellular transfer of radiolabelled metabolic intermediates (Subak-Sharpe, et. al., 1969; Pitts and Simms, 1977). Most systems have exhibited a sensitivity to changes in intracellular pH and/or free Ca^{++}, with a drop in the former, or a rise in the latter resulting in an uncoupling that is generally reversible upon return to normal values (Turin and Warner 1977; Rose, et. al., 1977; Spray, et. al., 1981). The ocular lens fiber cells of the chick, which will be discussed later, are a noteworthy exception, exhibiting reversible sensitivity only in the early stages of development. With differentiation, the reversibility of this sensitivity is lost (Scheutze and Goodenough, 1982; Miller and Goodenough, 1985).

In contrast to the wide base from which physiologic data have been drawn, biochemical and detailed structural data have been generated almost exclusively from studies on rodent hepatocyte gap junctions. The reasons for this are many. Historically, the gap junction, while observed in many tissues, was first experimentally defined in the rodent liver (Revel and Karnovsky, 1967). This, combined with initial observations of gap junction-like structures in the crude membrane preparations of subfractionated rodent livers (Benedetti and Emmelot, 1965), originally opened the door to further study in the rodent liver. With this as a starting point, protocols for the purification of rodent junctions eventually emerged (Benedetti and Emmelot, 1968; Goodenough and Stoeckenius, 1972; Evans and Gurd, 1972; Hertzberg and Gilula, 1979; Henderson, et. al., 1979).

Attention subsequently remained focussed on rodent livers in large part because purification of gap junctions by cell fractionation techniques is a very expensive, low-yield, high-risk protocol, and one which has proven not to be easily adaptable to other tissues, or even to the same tissue from other species (Kensler and Goodenough, 1980; Gros, et. al., 1983).

Great strides which have been made in the biochemical and ultrastructural analyses are principally due to two unique and inherent characteristics of rodent hepatocyte gap junctions: first, this structure has proven resistant to extraction in at least two detergents, N-lauryl sarcosine, and sodium deoxycholate (Benedetti and Emmelot, 1968; Evans and Gurd, 1972; Goodenough and Stoeckenius, 1972), permitting solubilization of the non-junctional membrane, and preservation of the intact junction. With this type of extraction included in the protocol, gap junctions have been purified to quite striking homogeneity, at least by morphological criteria (Hertzberg and Gilula, 1979). SDS polyacrylamide gel analysis (SDS PAGE) of purifed junctional preparations indicate that they are dominated by a single polypeptide with a relative molecular weight of about 27 kilodaltons. Other proteins are present, but in significantly reduced quantities (Henderson, et. al., 1979; Finbow, et. al., 1980). Subsequent immunocytochemistry on isolated hepatocyte membranes using antibodies to this 27 kd protein, have demonstrated immunoreactivity in the junctional domain, but not elsewhere (Janssen-Timmen, et. al., 1983).

The second inherent characteristic which has permitted an unparalled level of structural analysis is the tendancy for subunit particles of the junction to condense in the plane of the membrane, into relatively well-ordered hexagonal arrays. This paracrystallinity in two dimensions permits the application of computer-based image averaging techniques to electron microscope images of isoalted, negatively-stained junctions. The repetitive order, combined with the success at purififcation, also permits the application of X-ray diffraction. Thus, two high-resolution approaches which are generally not applicable to non-junctional membrane have been successfully employed. Several investigators have offered such analyses of the subunit structure of gap junctions (Zampighi and Robertson, 1973; Caspar, et. al., 1977; Zampighi, et. al., 1982; Unwin and Zampighi, 1980; Baker, et. al., 1983). Generally, these models describe the hepatocyte junction as a locus in the plasma membrane, usually on the order of 0.1 micron in diameter, of two adjacent cells where the extracellular space is narrowed and constant at about 2 nm. Concentrated at this site are transmembrane junctional subunits (connexons). Each connexon, in turn, can be thought of as two six-staved, transmembrane barrels, one in each plasma membrane, that meet end-to-end in the extracellular space, forming a conduit from the interior of one cell to the interior of the next. Each "stave" is identical, and believed to be comprised of a single 27 kilodalton polypeptide. Thus, the complete transmembrane channel has a total molecular weight of about 324kd (27 kd x 12).

More recently, the ocular lens has become a focus of attention in gap junction research. The principal reason for this is the extraordinary number of gap junction-like intercellular junctions that appear between the differentiating fiber cells of the lens (figure 1a). (Because a number of differences between "gap" junctions of fiber cells and the hepatocyte gap junction have been documented, the former will be referred to as "fiber cell junctions", to acknowledge the disimilarity.)

In order to understand the directions in which lens fiber cell junction research has gone, and to understand the potential importance of these junctions to lens biology, a brief review of the germane points of lens anatomy and development will be presented. (see reviews by Smelser, 1965, and McAvoy, 1981, for a more thorough treatment of the developing lens.)

In vertebrates, the lens begins its life as a thickening of head ectoderm overlying the presumptive eye cup (figure 2a). Under the influence of the eye cup this ectoderm invaginates, and eventually pinches off to form a hollow cyst of cells whose wall is a monolayer of epithelium. With time, the cells in the posterior half of the cyst elongate, gradually obliterating the lumen of the lens, and contacting the anterior epithelium (figure 2). At this point the lens has assumed its adult format (figures 2a-8 and 2b), a mass of cells elongated in the antero-postero direction, with a monolayer of epithelium on the anterior surface. The entire structure is enclosed in a thick collagenous capsule.

Growth in the diameter of the lens occurs throughout the life of the organism, and occurs in a unique fashion. New layers of fiber cells are continually added to the lens, as rings of epithelial cells in the coronal equator of the lens begin to differentiate. The differentiation process first manifests itself by the gradual elongation of the epithelial cell in the anterior and posterior directions. Anteriorly the elongating cell makes its way toward the anterior pole by inserting a process between the overlying epithelium and underlying fiber cells. Posteriorly the elongation proceeds between capsule and underlying fiber cell. Elongation continues until all the anteriorly- and posteriorly extending processes of a generation of cells meet at their respective poles, thus completing the addition of a new cell layer to the lens. As successive generations repeat this process, new layers are added upon old, and the lens increases in diameter. In such a fashion, the history of the lens is continually recorded, like the growth rings of a tree, with the epithelium acting as progenitors for the elongated fiber cells.

A result of this is that lens cells are arranged in a "linear" fashion according to developmental maturity. The least differentiated, most immature cells are found in the germinative zone, with maturational state progressing as one moves laterally through the epithelium to the equator, and then progressively from the surface toward the lens interior.

The cytologic changes which accompany this differentiaion are striking. The undifferentiated epithelial cell is an unremarkable cell, about 15 microns across, containing a complement of organelles expected of a metabolically moderately active cell; that is, a nucleus, mitochondria, endoplasmic reticulum, golgi, etc. With differentiation the cell elongates to several hundred times its original size, loses all of its membranous organelles, and, apparently, all of the functions associated with them (Wanko and Gavin, 1959; Cohen, 1965; Leeson, 1971; Rafferty and Esson, 1974; Kuwabara, 1975).

In contrast to this general loss of cytoplasmic detail, the plasma membrane assembles vast numbers of intercellular junctions which, superficially at least, resemble hepatocyte gap junctions. That is, they are small domains of adjacent cell plasma membranes, where the extracellular space is sharply narrowed, whose freeze fracture image consists of a clearly delimited collection of intramembrane particles.

The precise percentage of the membrane that is involved in these junctions is difficult to quantitate because the density of junctions is not uniform over the length of the cell (compare figures 1a and 1b). However, carefully-qualified estimates of the number of junctions present in the surface cortical cells of the chick lens run as high as 64%, some thirty-fold greater than found in the hepatocyte (Kuszak, et. al., 1978), agreeing with estimates drawn from isolated membranes (Lasser, et. al. 1972).

Biochemical changes which accompany differentiation are equally dramatic. The onset of differentiation is marked by the synthesis of tremendous levels of two classes of lens-specific proteins, widely accepted as markers of differentiation, the cytoplasmic crystallins, and the integral membrane protein, the Main Intrinsic Polypeptide (MIP) (see reviews: Harding and Dilley, 1976; Bloemendal, 1977; Bloemendal 1979). At maturity the MIP represents about 90% of the integral membrane protein of the lens (figure 3), and the crystallins about 30% of the wet weight of the lens (Benedetti, et., al., 1976).

Few cells undergo so dramatic a change in the course of differentiation as the lens cell. In those that do, such as neurons and muscle, we have been able to experimentally intervene, dissect the specialized functions at the cellular and molecular level, and understand the need for such profound change. In muscle, for example, the isolation of myofilaments, combined with the ability to biochemically dissect events in the process of filament sliding has led to a molecular understanding of the contractile process. The lens, however, has thus far largely defied the use of experimental intervention as a means of understanding its developmental changes. The principal reason for this impasse is that the function of the lens is so passive: it simply refracts light. What variation in its function that does occur, occurs passively at the hands of the ciliary body, which either exerts tension on the lens, changes its shape and thus its focal point, or relaxes the tension, allowing for passive recoil to its original shape. Thus, discussion of the changes that occur in the lens with differentiation has been largely confined to telelogic guesswork:

What are the functions of the crystallins, which are synthesized is such unique abundance? It has been postulated that the crystallins function to increase the index of refraction, thus ehancing the only function of the lens.

Why must the fiber cell lose its organelles, and, thus any semblance of self-sufficiency? It is assumed that the loss of

Figure 1a. Freeze fracture replica of ocular lens fiber cell. Image encompasses an entire fiber cell. Ridges at the arrows indicate the cell boundaries. Fiber cell junctions are present in extraordinary numbers at this location. One junction is demarcated by the bracket. Bar = 1.75 μ.
Figure 1b. From the same replica as in figure 1a, only the unequal distribution of fiber cell junctions is highlighted by the virtual absence of junctions in this field. The image, as in figure 1a, spans one complete fiber cell. A lone junction (arrow) can be seen. Bar = 0.7 μ.

1a

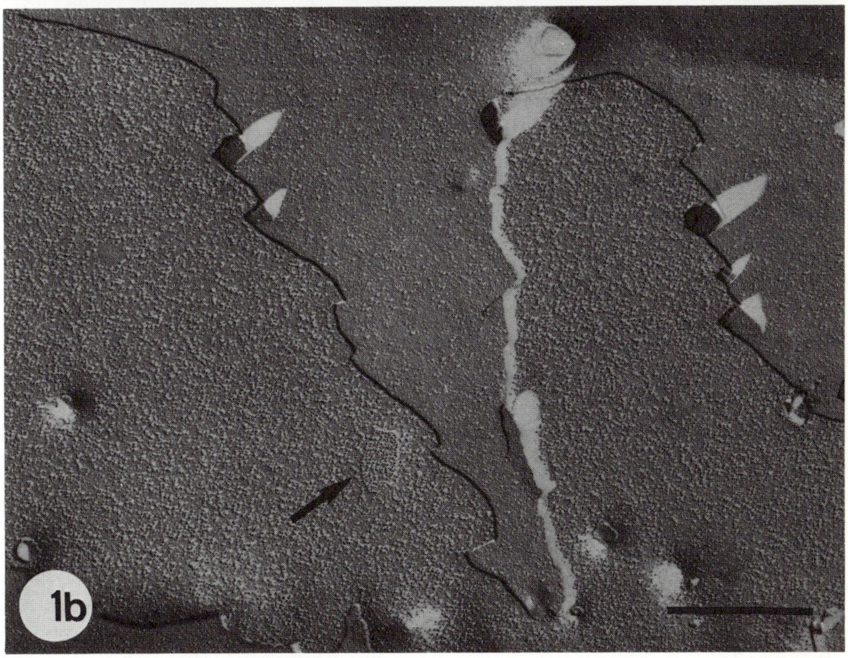
1b

organelles reduces light scatter, and minimizes selective light absorption due to the presence of chromophore-bearing enzymes.

Why is the number of junctions between fiber cells so unusually large? Speculation usually centers on the relative metabolic "helplessness" of the lens fiber cells, and their apparent reliance on the epithelium for energy-consuming, "house-keeping" functions. Experimentally this "helplessness" appears well-documented. Autoradiographic studies which monitored the incoporation of nucleotides and amino acids reveal that virtually no synthesis of either protein or nucleic acids is occurring in the mature fiber cells (Hanna, 1965; Young and Fulhorst, 1966). Studies on the uptake of metabolic precursors, such as amino acids and sugars, suggest that these transport processes are confined to the epithelium (Goodenough, et. al., 1980). Physiologic and cytochemical localization of the sodium-potassium ATPase suggest that it is confined to the epithelium, and first few layers of fiber cells (Bonting, et. al., 1963; Kinsey and Reddy, 1965; Duncan, 1969; Unakar and Tsui, 1980). This enzyme is largely responsible for the generation of the transmembrane electrochemical gradient that exists across cell plasma membranes, a function which is one of the largest energy expenditures a typical cell makes.

The question arises, then, as to how the mature fiber cell, lacking all organelles of intermediate metabolism, synthesis, and renewal, and lacking even the membrane-bound transport functions necessary for even the most basic of functions, can survive? For example, how can a fiber cell, in the absence of sodium-potassium ATPase, generate a transmembrane potential?

The answer appears to lie in the presence of an extraordinary number of intercellular junctions. It has been presumed, and experimental evidence supports this, that the fiber cell junction is analogous in function to the gap junction. That is, it permits the passive diffusion of ions and small metabolites between coupled cells (Rae, 1973, 1974, 1979; Scheutze and Goodenough, 1982; Rae and

Figure 2a. Schematic diagram depicting the development of the ocular lens. The ectoderm (d) thickens, invaginates, and ultimately pinches off to form a free-floating, hollow cyst of cells. With time, cells at the posterior of the presumptive lens elongate anteriorly, until finally contacting the anterior epithelium.
Figure 2b. Schematic diagram depicting the "format" of the mature vertebrate lens.

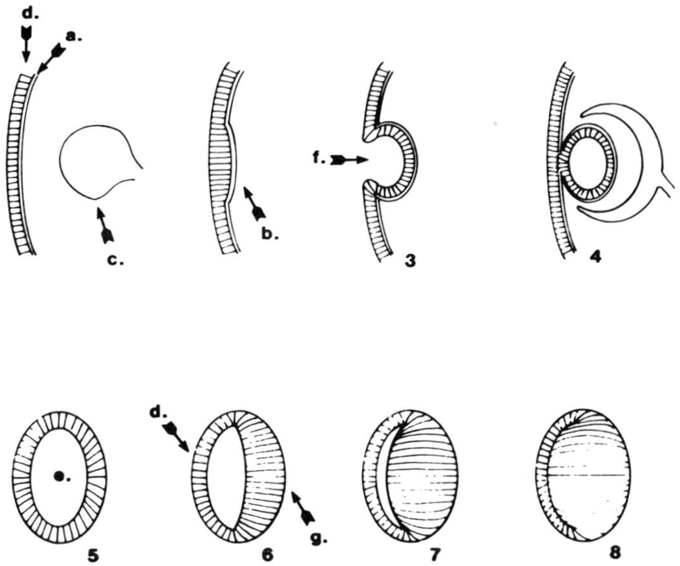

a. BASAL LAMINA
b. LENS PLACODE
c. OPTIC VESICLE
d. EPITHELIUM
e. LENS VESICLE
f. INVAGINATION
g. PRIMARY LENS FIBER CELL

2a

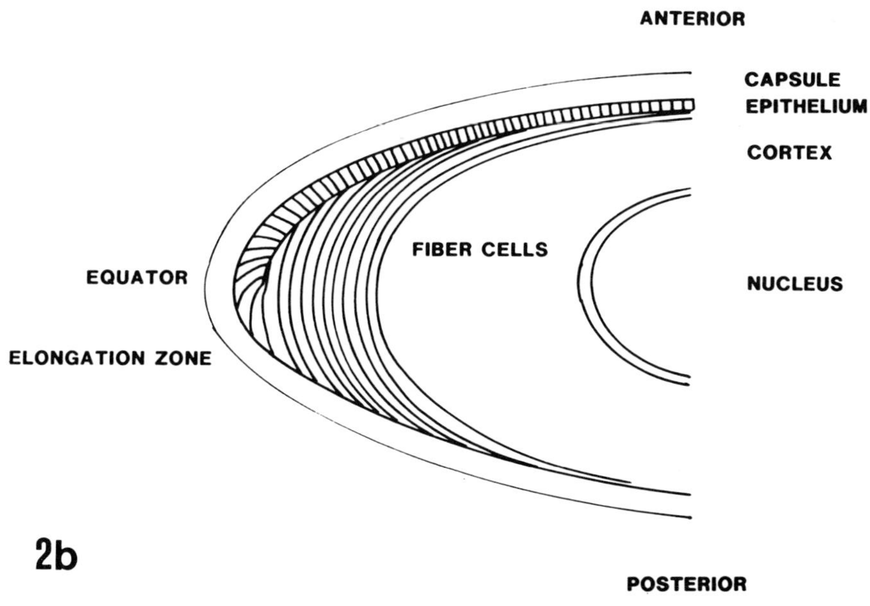

2b

Kuszak, 1983). In this fashion the acquisition of nutrients, the dissemination of wastes, and the maintenance of a transmembrane electrical potential, can all be maintained in the fiber cells, by the epithelial cells, via the intercellular junctions. In so doing, the number of cells that must have all the necessary, but optically undesirable metabolic machinery, are kept to a minimum.

From a functional standpoint, then, the lens presents itself as the most compelling example of the need for junction-mediated metabolic cooperation in non-excitable cells. While the impact of a failure in intercellular communication between hepatocytes is not clearly obvious, such a failure would clearly be catastrophic to the lens as an organ. The surface epithelial cells, having free access to the aqueous humor, would probably fare quite well, but the fiber cells, lacking both vasculature and ready access to the nutrients of the aqueous, would be in an uncommonly and perhaps fatally isolated position.

The attractiveness of the lens as a research subject for the study of plasma membranes, membrane proteins, and junction-mediated communication is further enhanced by nature of its development and anatomy. Specifically, because the lens lacks vasculature, innervation and cellular connective tissue, and because the lens is comprised almost exclusively of organelle-free fiber cells, it is extraordinarily simple to generate a uniquely clean and homogeneous plasma membrane fraction rich in intercellular junctions. (Bloemendal, et., al., 1972; Dunia, et. al., 1974; Benedetti, et. al., 1976; Goodenough, 1979; and figure 3, lane 1).

If preparations such as these are analysed by SDS polyacrylamide gel electrophoresis, they are seen to be dominated by the MIP. It is precisely this coincidental abundance of both the MIP and the fiber cell junctions which has led to extensive efforts to experimentally link the two. The fact that fiber cell junctions and hepatocyte gap junctions are morphologically similar, that the MIP and the hepatocyte junctional protein have comparable relative molecular weights, and that cells from a wide variety of tissues, including lens and liver, could form communicating junctions in vitro, (Hyde, et. al., 1969; Michalke and Lowenstein, 1971), even across species (Epstein and Gilula, 1977), led to transient speculation that gap junctions might be composed of a single well-conserved gene product, or at least a closely related family of genes. These observations have led to extensive efforts to

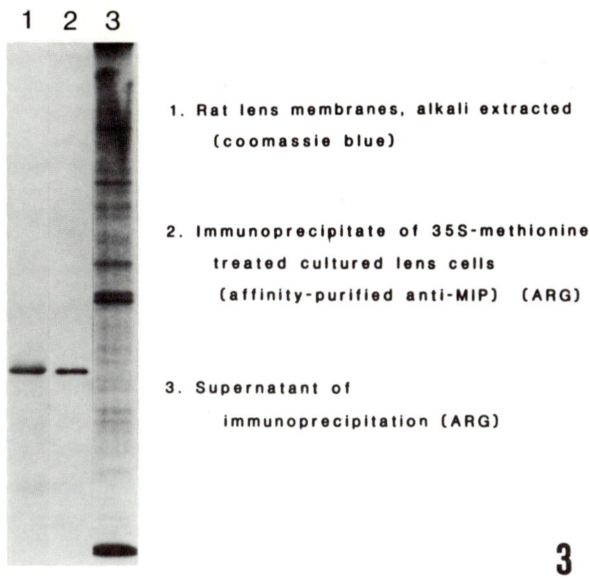

Figure 3.
Lane 1. A coomassie blue-stained lane from a 12.5% SDS polyacrylamide gel, on which has been resolved a sample of rat lens membranes prepared as according to Russel, et. al., 1981. The lane is deliberately underloaded to produce a standard for localizing in vitro synthesized MIP in lane 2. MIP so dominates the membrane protein profile that no other bands are even detectable at this loading.
Lane 2. Autoradiogram of immunoprecipitated MIP from 35S-methionine primed rat lens cultures. A membrane fraction was prepared from labelled cultures, solubilized in NP40, incubated with affinity-purified rabbit anti-bovine MIP, and harvested with Protein A-Sepharose 4-B (Pharmacia). The precipitate was washed, solubilized in cold SDS, and resolved on a 12.5% SDS polyacrylamide gel.
Lane 3. Same as in lane 2, but the supernatant from the immunoprecipitate was resolved. It can be seen that virtually all MIP was immunoprecipitated, and that due to the large ratio of epithelial cells to lentoid cells, MIP is not so dominant as in lane 1.

unequivocally link the MIP with the fiber cell junction, and to compare the fiber cell junction with the hepatocyte gap junction

(Goodenough, et. al., 1978; Goodenough, 1979; Hertzberg, 1980; Takemoto, et. al., 1981; Kuszak et. al., 1981; Hertzberg, et. al., 1982; Hertzberg and Gilula 1982; FitzGerald, et. al., 1983; Nicholson, et. al., 1983).

Many veterans of the struggle to purify hepatocyte gap junctions brought their expertise to bear on the purification of fiber cell junctions (Benedetti, et. al., 1976; Goodenough, 1979). However, fiber cell junctions proved to be much less resistant to deoxycholate and sarcosine extraction, and thus defied purification to the same degree of morphological homogeneity as the hepatocyte gap junction. Nevertheless, although not "pure", the fiber cell membrane preparation appeared to be enriched in fiber cell junctions, and the SDS PAGE analysis suggested an apparent co-enrichment of the MIP (Goodenough, 1979). However, while MIP certainly dominated such preparations, several other polypeptides were present as well. Thus, the inability to achieve homogeneity in the fiber cell junction preparation precluded any unequivocal correlation between a morphologically "pure" preparation and a biochemically "pure" gel, such as has been achieved for hepatocyte junctions. Nevertheless, the coincidental abundance of both the fiber cell junction and the MIP certainly made MIP a leading candidate for the structural protein in these junctions.

The fiber cell junctions have also defied the application of high-resolution structural analysis. Such analyses are performed on negatively-stained, isolated membranes, in which the connexons have condensed into domains of hexagonally-packed arrays. However, fiber junctions have not demonstrated the capacity to condense into ordered arrays in such preparations (Goodenough, 1979; Zamphighi, et al., 1982), thus eliminating the possibility of applying averaging techniques, or x-ray diffraction. A number of particle-packing patterns have been described in freeze-fracture replicas of lens membranes, under a variety of conditions, but whether these are related to the lens junctions is not clear (Peracchia and Peracchia, 1980 a, b; Zampighi, et. al., 1982; Kistler and Bullivant, 1980)

Immunocytochemical technology was at the heart of the next major effort applied in linking MIP with the fiber cell junction. Immunofluorescence studies, using anti-MIP on frozen, thick (7 micron) sections, however, failed to clarify the situation. While these studies yielded valuable information concerning the timing of appearance and the overall distribution of the MIP, they did not shed further light on the role of the MIP in intercellular junctions (Waggoner and Maisel, 1978; Broekhuyse, et. al., 1979)). The observation that MIP labelling was uniform throughout the fiber cell membrane, and not punctate, as one might expect if only junctions had labelled, is not necessarily inconsistent with the thought of MIP as a junctional protein. The relatively small size and high density of the junctions, compared to the thickness of the sections, would allow for extensive overlap in the viewing axis, conveying the impression of homogeneous membrane labelling. Similarly, the observation that the lens epithelium lacked MIP immunoreactivity was not inconsistent with the role of MIP in lens junctions, either. The lens epithelium is very poor in intercellular junctions, with something on the order of 1% of its surface area involved in gap junctions. Further, what junctions the lens epithelium does have are located at the apico-lateral corner of the cell, very close to, and in frozen thick sections, indistinguishable from the fiber cell plasma membrane. Thus, interepithelial cell gap junctions might well have been MIP-positive, but their small number and proximity to the MIP-rich fiber cell would have masked their presence.

Bok et al. (1982) carried the MIP immunocytochemical localization to the electron microscope level using antiserum to the MIP, and isolated lens fiber cell membranes. This preparation appeared to be almost exclusively junctional, and was heavily labelled, suggesting extensive presence of MIP in the fiber cell junction. These investigators also noted that the occasional unit, or non-junctional membrane was also labelled, suggesting the non-junctional presence of MIP as well. However, these investigators made no definitive statement regarding such labelling because it was unclear whether these unit membranes were actually junctional membranes split in the course of the isolation, or originally non-junctional membranes.

Zampighi et. al. (1982), using a different isolation protocol, and Paul and Goodenough (1983), using the original Goodenough isolation, both introduced the use of tannic acid as a fixative for lens membranes. The inclusion of tannic acid resulted in the delineation of two distinct subpopulations of pentalaminar, junction-like structures, in a preparation once considered to be comprised almost exclusively of a single class of junction. The two junction-like structures differ markedly in the overall thickness of the junctional profile, distinguished by significant differences in their overall thickness (12 nm vs. 16 nm).

The use of tannic acid in the Paul and Goodenough study also produced an apparent increase in the preservation of unit membranes, revealing that the original isolation protocol was less of a junction-enrichment than previously believed. These investigators, in performing anti-MIP immunolabeling in combination with tannic acid fixation, found MIP immunoreactivity only in the non-junctional membrane, and in the thinner of the two sets of pentalaminar profiles.

It remains, at present, unclear what these two populations of junction-like profiles represent, and what their relationship is to the fiber cell junction. Rafferty and Esson (1974) have documented the existence of at least two, and possibly three subpopulations of lens junctions in situ, also discriminable on the basis of thickness, with thicker junctions occurring between epithelial cells, and thinner junctions between fiber cells. However, because Zampighi, et. al., and Paul and Goodenough decapsulated the lenses before homogenization, (thereby removing the epithelium) it seems unlikely that the thicker profiles they observed are epithelial in origin. An alternate explanantion for the presence of two junctional populations in the Zampighi, et. al. and Paul and Goodenough preparations, is that two unique classes of junctions are present in the fiber cell (one perhaps a carry-over from the epithelium) which have gone unnoticed due to the subtlety of the difference between the two. A second interpretation is that the two subclasses are actually the same junctional type, but in two different states. Finally it has been suggested that only one of the subclasses is actually junctional, and that the second pentalaminar profile is simply the artifactual adhesion of two adjacent plasma membranes which occurs during the long and harsh isolation, mimicing a junction in cross-section. This latter explanation, however, does not account for the observation of Rafferty and Esson, of two or three distinct junctional types in situ.

What is clear from the enhanced fixation of tannic acid, however, is that a great deal more non-junctional membrane is present in these preparations than was formerly thought.

The interpretive problems raised in these studies have also underscored the limitations in our ability to extrapolate from studies on isolated membranes. Several potentially valid criticisms will always limit the generality of the results: first, the isolation and fractionation procedures typically used in these protocols may well result in the artifactual enrichment for some unrepresentative subclass of lens membranes or junctions. Second, the potential for redistribution of membrane components during the extraction in urea, flotation on a gradient, and the lengthy immunolabelling, which can extend the procedure to 48 hours or more, cannot be discounted. Finally, the in vivo spatial relationships are destroyed.

In an effort to circumvent problems of this type FitzGerald et. al. (1983) immunocytochemically localized the MIP at the electron microscope level, in ultrathin frozen sections of fixed lens. Their results indicated that MIP was present in both junctional and non-junctional domains, with no detectable difference between the two. This finding meant that, by its distribution at least, MIP could not be considered a junctional protein, but simply a membrane protein with no differential distribution. Such a finding does not preclude the possibility that MIP is a junctional protein, however. MIP may well exist in some form of physiologic equilibrium between junctional and non- junctional domains, such as a product-precursor relationship. However, the result was that, once again, experimental efforts had failed to cement the relationship between the MIP and the fiber cell junction.

In an effort to define the relationship of MIP to the hepatocyte gap junction, these investigators extended the MIP immunocytochemistry to frozen thin sections of rat liver. Their results, which showed no sign of MIP presence in the hepatocyte junction, supported biochemical and immunochemical evidence from several sources which were underscoring the lack of homology between the MIP and the hepatocyte junctional protein (Hertzberg, 1980; Nicholson, et. al., 1983). Partial amino acid analysis, peptide mapping, and assays for immunologic cross-reactivity all suggested that the two proteins had little or no sequence homology. Subsequent probing of liver mRNA with cDNA probes derived from the cloned MIP gene (Gorin, et. al., 1984), as well as immunocytochemistry of the lens with anti-liver junctional protein, have confirmed this apparent lack of homology.

Differences between MIP and the hepatocyte junctional protein would not be surprising, of course if MIP eventually proved to be unrelated to fiber cell junctions. However, studies which examine the fiber cell junction itself, without regard to its composition, have resulted in a lengthening list of structural and functional differences between it and other well-studied gap junctions: 1) the fiber cell junction lacks an easily demonstrable "gap" between the outer leaflets of the contributing cell membranes (Bloemendal, et. al., 1972; Phillipson, et. al., 1975; Friedlander, 1980; Hertzberg, et. al., 1982). This is in contrast to other tissues such as heart and liver where the 2 nm gap is readily visualized (Revel and Karnovshy, 1967; Goodenough, et. al., 1978; Kensler and Goodenough, 1980). 2) In situ, the intramembrane particles in freeze-fracture replicas of fiber cell junctions are rarely seen to condense into the hexagonal arrays so commonly seen in other tissues (Chalcroft and Bullivant, 1970; Goodenough and Revel, 1970; Bloemendal, et. al., 1972; Benedetti, et. al., 1974; Peracchia, 1977; Kuszak, et. al., 1978; Goodenough, 1979; Friedlander, 1980; and figures 4a,b). Peracchia has, however, demonstrated such packing in vitro, on isolated lens membranes. The role of this condensation in normal junctional physiology, if any, is not understood (Raviola, et. al., 1980; Hanna, et. al., 1984). 3) Most gap junctions which have been studied physiologically have demonstrated a capacity for reversible assumption of a high-resistance state in response to stimuli such as pH drop or Ca^{++} rise. In contrast, the junctions of chick lens fiber cells have demonstrated no such reversibility. While they can be uncoupled, the closure is not reversible, at least under conditions that restore coupling in other systems (Scheutze and Goodenough, 1982; Miller and Goodenough, 1985). Frog fiber cell junctions, however, have demonstrated coupling recovery after dinitrophenol uncoupling (Rae, et. al., 1982). 4) finally antibodies to hepatocyte gap junction protein, which label just the

Figure 4a. Freeze fracture replica of a "fiber cell" junction from between two lentoid cells in rat lens cultures. This particular junction typifies the loosely arranged, unordered packing of fiber cell junctions. Bar = 0.04 μ

Figure 4b. Freeze fracture replica of a gap junction from between two undifferentiated cells in rat lens culture. This junction shows micro-domains of much more condensed and ordered packing (circle) of intramembrane particles (IMPs) that typifies the junctions between lens epithelial cells in vivo, and junctions from most other tissues. The degree of IMP packing in this image is relatively minimal compared to that observed routinely in hepatocyte and other junctional types. Bar = 0.05 μ

junctional domain, and not non-junctional membrane, do not label lens fiber cells.

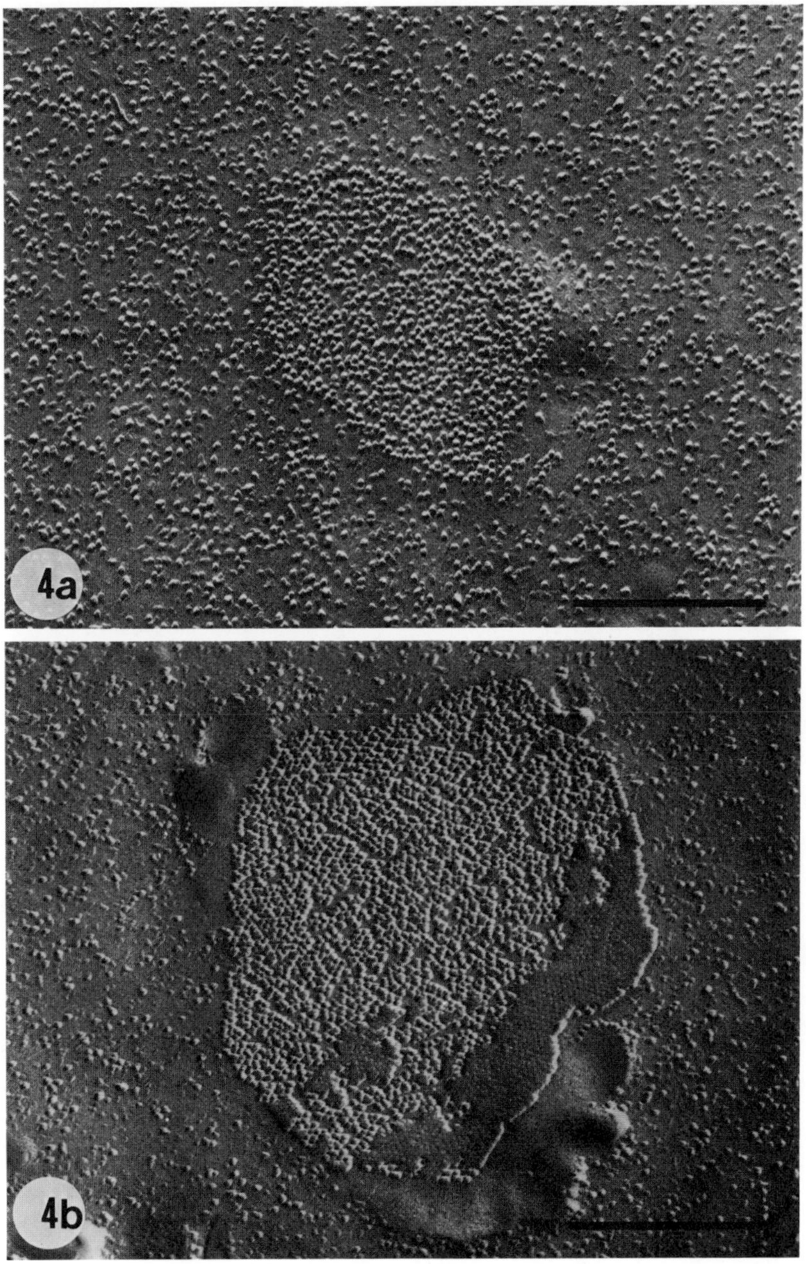
4a
4b

Thus, in light of the large number of structural, immunocytochemical, and physiological differences which exist between the fiber cell junction and other gap junctions, particularly the hepatocyte junction, it would not be unexpected that they would exhibit compositional differences as well.

Immunocytochemical, immunochemical, molecular biological, and biochemical approaches have all suggested, then, that the MIP and the hepatocyte junctional proteins are at least substantially different, and possibly unique polypeptides. They have in common only a similar, but not identical weight (Hertzberg, et. al., 1982), and possibly analogous functions. One report, however, has emerged which suggests some degree of immunologic cross-reactivity between the two polypeptides (Traub and Willecke, 1982).

Even if one assumes that the MIP is not the fiber cell junctional protein, one is inclined to surmise, but with somewhat less confidence, that hepatocyte junctions and fiber cell junctions, while possibly analogous in function, are certainly not identical, at least morphologically and immunocytochemically.

To further complicate the issue, there is a growing body of evidence suggesting that the "gap" junctions between lens epithelial cells may well be different from those between fiber cells: 1) freeze-fracture images of epithelial cell junctions and fiber cell junctions routinely differ, with the intramembrane particles of the epithelial cell junctions exhibiting a tendency to condense, and those of the fiber cell junctions remaining dispersed (Bloemendal, et., al., 1972; Benedetti, et. al., 1974; Peracchia, 1978; Goodenough, 1979; Kuszak, et. al., 1982; figures 4a,b); 2) thin section ultrastructural examination of the lens reveals differences in the thicknesses of the junctions that exist between epithelial cells, and those that occur between fiber cells (Rafferty and Esson, 1974); 3) physiological studies indicate that differences exist in the sensitivity of the two populations to uncoupling procedures (Scheutze and Goodenough, 1982; Miller and Goodenough 1985); 4) finally, the immunolabelling of frozen thin sections revealed that MIP was not present in the junctional complexes between epithelial cells (FitzGerald, et. al., 1983).

Thus, some of the morphological differences which distinguish hepatocyte and fiber cell junctions, also distinguish lens epithelial cell and lens fiber cell junctions. Further, if the MIP indeed represents the fiber cell junctional protein, then the immunocytochemistry, which shows no MIP immunoreactivity in epithelial cell junctions, provides convincing evidence that two classes of junctions are synthesized by the lens in the course of development: an MIP-positive class between fiber cells, and an MIP-negative class between epithelial cells.

It appears then, that enough physiological, morphological and immunocytochemical evidence exists to preclude assuming that lens epithelial cell junctions and lens fiber cell junctions are the same.

Clearly a great deal would be clarified if the role of the MIP were unequivocally defined. It has been widely assumed that the MIP is the fiber cell junctional protein, and, without question, circumstances make it the leading candidate. However, the evidence doesn't yet support the dogmatic acceptance of this as fact. The evidence is strictly circumstanatial, and in all cases open to alternate interpetation.

The most compelling and often-used argument in support of the MIP as a junctional polypeptide, is the coincidental abundance, and the apparent co-purification of MIP and fiber cell junctions (Goodenough, 1979). Stated in other words, it is difficult to imagine that one of the minor, non-MIP proteins in the SDS PAGE profile could account for the vast number of junctions present. Certainly the "coincidental abundance" line of thought is intuitively appealing, but it lacks the quantitiation that would qualify it as hard evidence. For instance, one could argue that it is not a matter of non-MIP proteins being too scarce to account for the large number of junctions, but, conversely, that MIP is too abundant for unknown purposes. That is, other proteins are present in sufficient quantity to account for the junctions, but MIP is present in freakish excess.

An important assumption in the "coincidental abundance" argument is that the "junctions" observed in the isolated membrane preparation are all in fact junctions, and importantly, all the same type of junction. The results of Zampighi, et. al. (1982) and Paul and Goodenough (1983), seriously undermine this assumption. Their results indicate that a preparation of lens membranes once thought to be dominated by a single class of junctions, with a smattering of unit membranes mixed in, is actually quite rich in unit membranes, and contains two morphologically discriminable classes of "junctions". Further, the immunocytochemical localization of Paul and Goodenough demonstrates a clear difference in immunoreacrtivity, suggesting a difference in composition. While arguments involving conformational changes in the junctional protein, resulting in alterations in available antigenic determinants can be invoked to explain the differences in immunoreactivity of these two populations, no concrete data exists to resolve the issue.

Thus, while the nature of these two classes of pentalaminar profiles is not known, their mere existence raises the possibility that the actual number of genuine junctions in this preparation may be quite a bit lower than expected, meaning that junctions are not as coincidentally abundant as thought, and increasing the odds that a minor, non-MIP proteins may conceivably account for them.

Another line of evidence that is frequently marshalled to support the role of the MIP as a junctional protein is the immunocytochemcal localization of the MIP in the domain of the junction (Bok, et. al., 1982; FitzGerald, et. al., 1983). Its presence within the junction , however must be tempered by its presence in non-junctional membrane as well. One can interpret these data in either of two ways: MIP is the fiber cell junctional protein, but it is in some sort of equilibrium with non-junctional MIP; or, MIP is not a junctional protein at all, but simply a general membrane protein that is not actively excluded from the domain of the junction. In either case, regardless of function, both junctional and non-junctional domains would be MIP-immunoreactive. It should be reiterated at this juncture, that immunolabelling of hepatocyte membranes with anti-liver junctional protein labels only the junctional domain, and not the adjacent plasma membrane (Janssen-Timmen, et. al., 1983).

It was becoming clear that the role of the MIP in lens fiber cell coupling needed to be approached from a different vantage point. One very promising avenue that is currently being explored in a number of laboratories is the incorporation of MIP into artificial membranes (Nikaido and Rosenberg, 1984; Peracchia and Girsch, 1984). The latter authors report that this approach results in the formation of channels with gap junction-like properties, with a sensitivity to calcium-binding proteins.

Alternatively we have set out to determine whether a mammalian lens culture system could be established that would differentiate sufficiently to express the MIP, and allow for an in vitro approach to the question of MIP's role.

Differentiation of lens cells in culture has been reported in a wide variety of systems, including rodent (Okada, et. al., 1971; Piatigorski, et. al., 1972; Mousa and Trevithick, 1977; Tsuenematsu, et. al., 1978; Beebe and Feagan, 1981; Counis, et. al., 1982). However, while these systems have shown extensive evidence of morphological and biochemical differentiation, including synthesis of the lens-specific crystallins, and the loss of organelles, the synthesis of MIP in mammalian culture systems has not been documented.

We have recently established a very simple system for culturing rat lens cells that result in quite extensive morphological and biochemical differentiation, including the expression of the MIP (FitzGerald, et. al., 1984). Details of the culturing process are being reported elsewhere (FitzGerald, et. al., 1985). In brief, the procedure consists of enzymatically "scrubbing" young rat lenses free of adherent debris, dispersal of lens cells by proteolysis and mechanical disruption, and seeding at high density on collagen-coated substrates. After twenty-four hours the non-adherent cells are washed away, and the cultures fed. Differentiation proceeds without further input over the course of two to four weeks.

The differentiation process is clearly evident with phase or Nomarski optics. Initially the cultures are comprised of isolated epithelial-like cells (that is, polygonal-shaped cells, with organelles). With time the cells multiply, approach confluence, and begin to segregate into clumps. These clumps begin to grow in the vertical axis, assembling into classical "lentoids" consisting of rounded mounds of greatly enlarged cells (figures 5, 6a). Confluence receeds around these mounds, ultimately leaving the culture dish spotted with lentoids which are perched on an underlying layer of epithelial-like cells, surrounded by bare dish.

Ultrastructurally the cells bear a striking resemblance to the bow region of the lens, with undifferentiated epithelial cells on the collagen-coated substrate (figure 7a), and a linear progression of increasingly differentiated cells as one rises through the lentoid. A progressive loss of organelles is observed, and the cytoplasm begins to fill with a general undetailed density (figure 7b), similar to that of the crystallin-rich fiber cell. The principal distinctions between differentiated cells in culture and those in vivo were: a) the differentiated cells in vitro lacked the

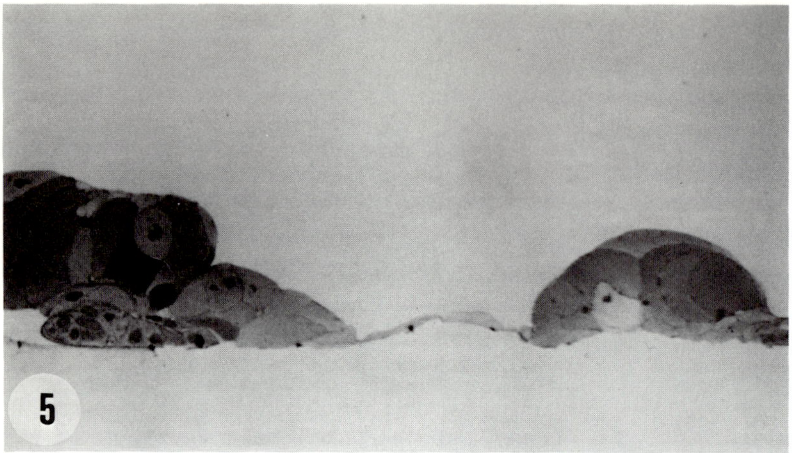

Figure 5. Light micrograph of a section through plastic-embedded rat lens culture. Two mounds of cells (lentoids) are obvious, connected by a a very flat monolayer of undifferentiated epithelial cells. The relative loss of cytoplasmic detail is apparent even at the light microscope level.

regular and ordered geometry of fiber cells in situ; and b) the fiber cells lacked the extraordinary numbers of junctions routinely found between fiber cells in vivo. Junctions were present, but at a frequency that was comparable to the level of inter-epithelial cell junctions.

To biochemically characterize the course of differentiation cultures were incubated in the presence of 35S-methionine, and the cell cytosol immunoprecipitated with antisera against alpha-, beta-, and gamma crystallins. The insoluble fraction was extracted with the detergent NP-40, and immunoprecipitated with affinity-purified anti-MIP. The immunoprecipitates were resolved by SDS PAGE and visualized by autoradiography.

It was evident from this analysis, even using unfractionated antiserum, that the crystallins were being synthesized. Similarly, it was documented that the MIP was being expressed (figure 3). Subsequent immunofluorescence studies, using the same antibody preparations revealed that this de novo synthesis of biochemical markers of differentiation was confined to the cells of the lentoid (figure 6a,c).

Having thus documented both morphological and biochemical differentiation we were satisfied that the cultures met minimum criteria for initiating studies into the role of the MIP and intercellular communication. The next step was to characterize the communication competence of these cells, and establish baselines for further experimental interventions.

The simplest approach for documenting intercellular communication in such a system is the iontophoresis of the fluorescent dye Lucifer Yellow CH using intracellular microelectrodes. The dye is readily visible and is small enough to spread via gap junctions to cells that are coupled to the injected cell.

When the primary injection site was a morphologically undifferentiated epithelial-type cell the intercellular spread of fluorescence was unrestricted and rapid, a result that is typical of gap junction-coupled cells (figures 6a,b). However, if the epithelial cell was adjacent to a lentoid, the dye spread in all directions, except into the lentoid, suggesting some sort of barrier between lentoid and epithelium. (figures 6a-c)

Conversely, when the site of injection was a differentiated cell of the lentoid, no spread of fluorescence was observed, suggesting that the lentoid cells were uncoupled, or did not possess junctions (figures 8a-c). This was a surprising result for a number of reasons. Most cells synthesize gap junctions, even in vitro; the lentoids were clearly rich in the MIP, a putative junctional protein; the differentiated fiber cell is an extraordinarily well-coupled cell in vivo; finally, ultrastructural examination of the lentoid cells documented the presence of junctions in numbers that, while low for a fiber cell, were not unusual for other cell types.

To clarify this, we re-examined the communication between cultured cells using the more sensitive criteria of electrical coupling, as measured with two intracellular electrodes, in separate cells. This revealed that the cells of the lentoid were, in fact, electrically coupled, both to each other, and to nearby epithelial cells.

At the moment we are unable to unequivocally explain the lack of dye coupling, but the presence of electrical coupling, between cells of the lentoid. One possible explanation is that the free diffusion of Lucifer Yellow CH is being restricted by charge interaction with the crystallins, and that the enormous levels of crystallins in the fiber cell serve as an effective buffer, tying up the Lucifer Yellow and preventing its spread. However, similar studies in vivo, in which fiber cells of the developing chick lenses and mature frog lenses have been injected, show no such restriction. Thus, this would not appear to be a viable explanation for restriction of dye spread among lentoid cells.

A second explanation is that there are far fewer junctions between lentoid cells than between epithelial cells, and that communication between lentoid cells is below the limit of detection of the relatively insensitive dye coupling, but well within the limits of the extremely sensitive electrical coupling. Our initial non-quantitative examination of thin sections and freeze fracture images, however, would not support this.

The most tantalizing explanation would be that we are dealing with two distinct classes of junctions, with different conductance properties. To accept this, however, one would have to argue that the fiber cell junction is capable of passing dye, but at a much reduced rate. Thus observed dye coupling between fiber cells in vivo (Rae, 1974; Scheutze and Goodenough, 1982; Rae and Kuszak, 1983) would be due to the presence of enormous numbers of junctions, and, conversely, the apparent lack of dye coupling in vitro is a matter of undetectability due to the small number of junctions.

Figure 6a. Rat lens culture, phase optics. In the plane of focus, at the center of the image, is a lentoid. Out of the plane of focus is a large area of undifferentiated lens epithelial cells. At the lower and upper right corners are areas of bare dish.
Figure 6b. Same field of view as in figure 6a, but photographed with fluorescence optics, to highlight the presence of Lucifer Yellow CH, which was iontophoresed into the very bright, undifferentiated cell at the center of the image. The fluorescence has spread to numerous neighboring undifferentiated cells, but not into the overlying lentoid, particularly evident at the site where the lentoid overlies bare dish (arrow).
Figure 6c. Same field as in 6a-b, but photographed with fluorescence optics designed to emphasize Rhodamine fluorescence. The cultures, after injection with Lucifer Yellow CH, were fixed, and immunolabelled with affinity-purified anti-MIP, and Rhodamine-conjugated goat anti-rabbit IgG. It is obviuos that MIP immunoreactivity is confined to the lentoid, confirming the morphological suggestion of differentiation.

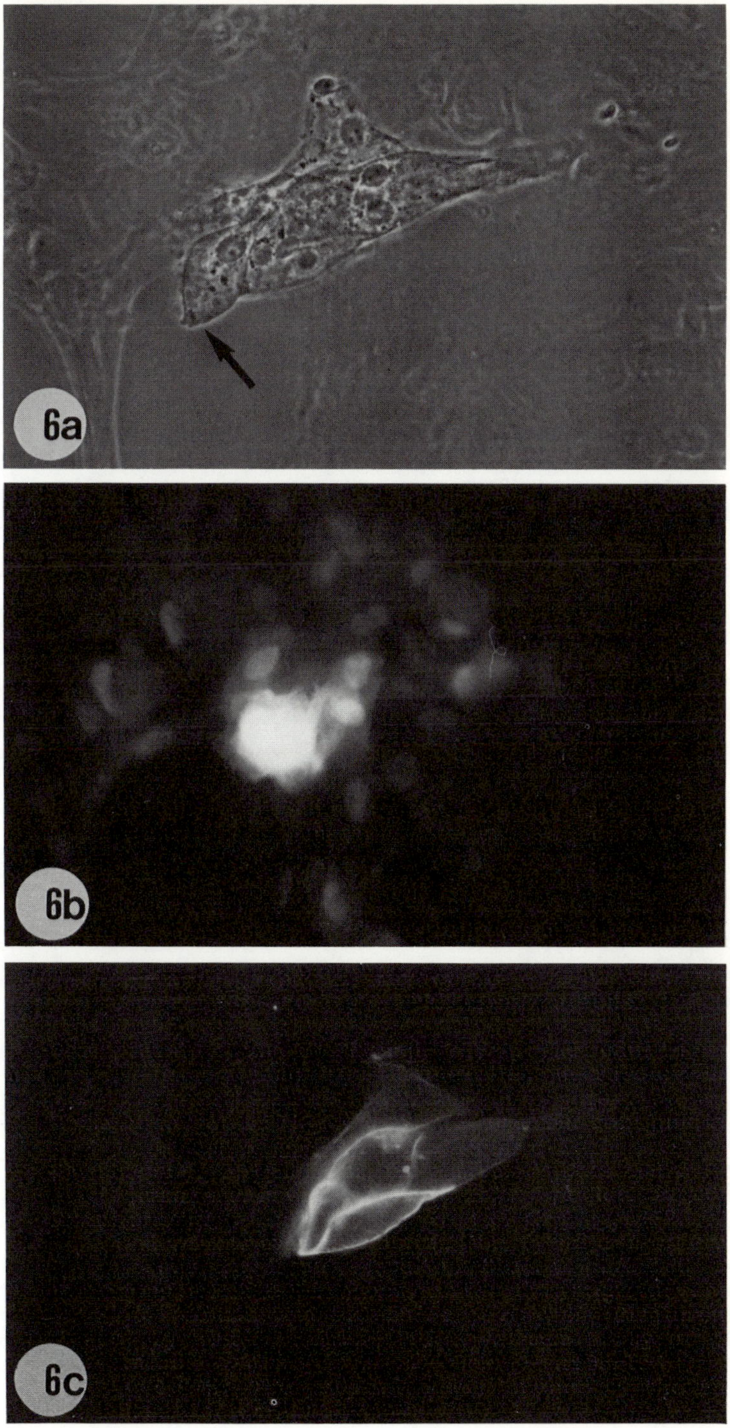
6a
6b
6c

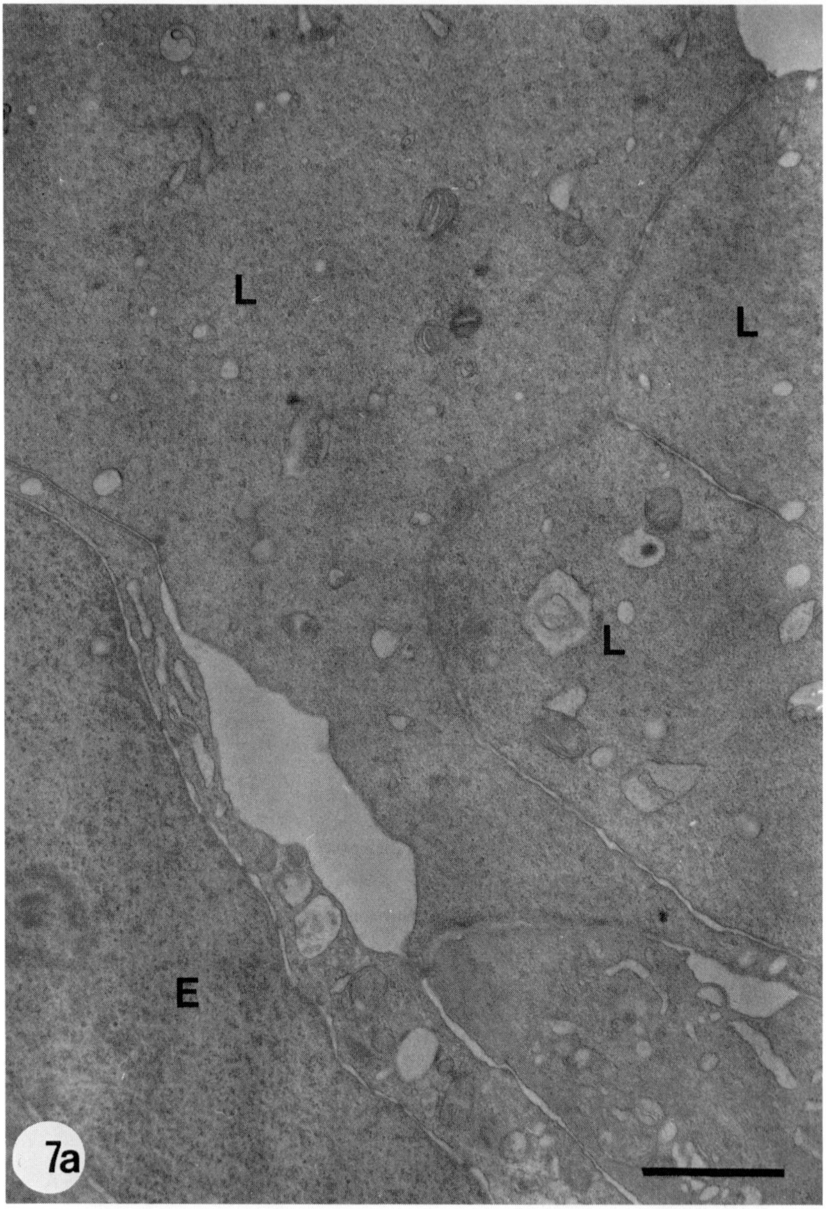

Figure 7a. Electron micrograph rat lens cells in culture. The field encompasses an undifferentiated, organelle-rich epithelial cell (E) which is attached to the collagen-coated dish. Overlying, partially- differentiated, organelle-poor lentoid cells (L) are seen toward the upper right. Bar = 0.5 μ

Figure 7b. Electron micrograph of rat lens cells in culture. The field encompasses four quite differentiated lentoid cells. Note the relative absence of intracellular organelles which typifies the differentiated fiber cell in vivo. The dish surface, and underlying epithelial cell, lies out of view to the bottom, and the upper surface of the lentoid is apparent at the top left of the image. Bar = 1 μ

Such disparity in the permeability properties of gap junctions has been documented in several systems (Lo and Gilula, 1979; Warner and Lawrence, 1982; Weir and Lo, 1982), however, the mechanism and interpetation remain to be elucidated.

Clearly the most elegant deployment of such a culture system would be the microelectrode-injection of antibodies to the MIP, in an effort to disrupt the elelectrical coupling between cells of the lentoid. However, the number of possible explanations for false negative effects renders the problem far less tractable than might be imagined.

For now, we will have to be content with the knowledge that the mere presence of MIP in the lentoid cell membrane is not, in itself, sufficient for the assembly of large numbers of fiber cell junctions, or the conferral of extensive dye-coupling properties.

Figure 8a. Rat lens cell culture, phase optics. The field of view contains a lentoid, above the plane of the bare dish.
Figure 8b. Same field as in figure 8a, but photographed with fluorescence optics to enhance Lucifer Yellow CH. In contrast to figure 6b, in which an undifferentiated cell was the site of microelectrode insertion, this injection was into a differentiated cell of a lentoid. Note that the dye has failed to spread to adjacent cells.
Figure 8c. Same field as in figure 8a-b, but photographed with fluorescence optics designed to enhance Rhodamine fluorescence. Culture was immunolabelled as described as described in figure 6c, with anti-MIP.

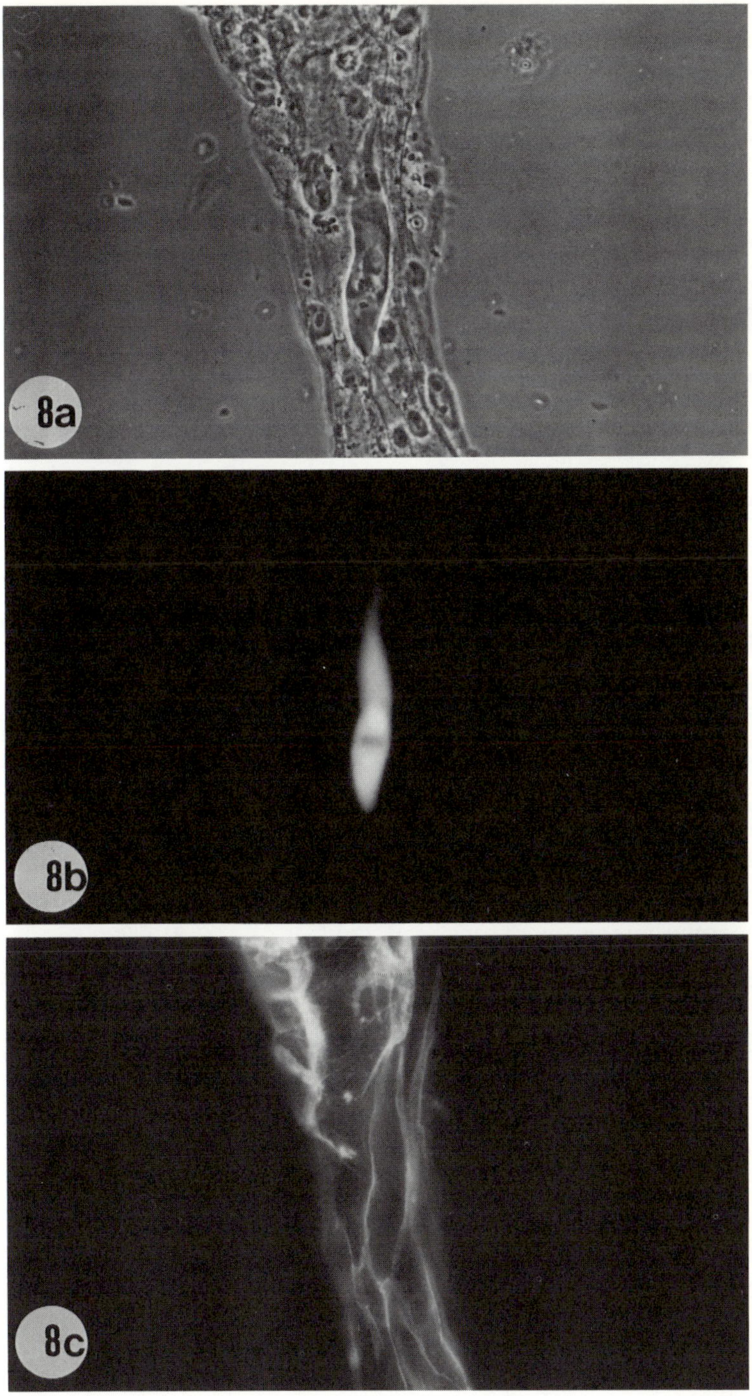
8a
8b
8c

REFERENCES

Asada, Y., and M. Bennett. 1971. Experimental alterations of coupling resistance at an electrotonic synapse. J. Cell Biol. 49:159-172

Baker, T., D. Caspar. C. Hollinshead and D. Goodenough. 1983. Gap junction structures. IV asymmetric features revealed by low-dose electron microscopy. J. Cell Biol. 96:204-216

Barr, L., M. Dewey and W. Berger. 1965. Propagation of action potentials and the structure of the nexus in cardiac muscle. J. Gen. Physiol. 48:797-823

Beebe, D., and D. Feagans. 1981. A tissue culture system for studying lens cell differentiation. Vision Res. 21:113-118

Benedetti, E., I. Dunia, C. Bentzel, A. Vermorken, M. Kibbelaar and H. Bloemendal. 1976. A portrait of plasma membrane specializations in eye lens epithelium and fibers. Biochim. et Biophys. Acta 457:353-384

Benedetti, E., I. Dunia and H. Bloemendal. 1974. Proc. Natl. Acad. Sci. 71:5073-5077

Benedetti, E., and P. Emmelot. 1968. Hexagonal array of subunits in tight junctions separated from isolated rat liver plasma membranes. J. Cell Biol. 38:15-24

Bennedetti, E., and P. Emmelot. 1965. Electron microscope observations on negatively stained plasma membranes isolated from rat liver. J. Cell Biol. 26:299

Bennett, M., and D. Goodenough. 1978. Neurosciences Res. Program Bull. 16:443

Bennett, M., E. Aljure, Y. Nakajima, and G. Pappas. 1963 Electrotonic junctions between teleost spinal neurons: Electrophysiology and Ultrastructure. Science 141:262-264

Bloemendal, H. 1977. The vertebrate eye lens. Science 197:127-138

Bloemendal, H. 1979. Lens proteins as markers of terminal differentiation. Ophthalmic Res. 11:243-253

Bloemendal, H., A. Zweers, F. Vermorken, I. Dunia and E. Benedetti. 1972. The plasma membranes of eye lens fiber. Biochemical and structural characterization. Cell Differentiation 1:91-106

Bok, D., J. Dockstader and J. Horwitz. 1982. Immunocytochemical localization of the lens main intrinsic polypeptide (MIP26) in communicating junctions. J. Cell Biol. 92:213-220

Bonting, L., L. Caravaggio and N. Hawkins. 1963. Studies on sodium potasium activated ATPase. Its role in cation transport in the lens of cat, calf, and rabbit. Arch. Biochem. Biophys. 101:47-55

Broekhuyse, R., E. Kuhlman and H. Winkens. 1979. Lens membranes VII. MIP is an immunologically specific component of lens fiber cell membranes and is identical with 26kd band protein. Exp. Eye Res. 29:303-313

Caspar, D., D. Goodenough, L. Makowski and W. Phillips. 1977. Gap junction structure I. Correlated electron microscopy and X-ray diffraction. J. Cell Biol. 74:605-628

Chalcroft, J. and S. Bullivant. 1970. An interpretation of liver cell membrane and junction structure based on observation of freeze fracture replicas of both sides of the fracture. J. Cell Biol. 47:49

Cohen, A. 1965. The electron microscopy of the normal human lens. Invest. Ophthal. Vis. Sci. 4:433-446

Counis, M., E. Chaudun, Y. Courtois, J. Carreau, J. Jack and R. Clayton. 1982. Differential protein synthesis of chick embryonic lenses during in vitro differentiation. Exp. Eye Res. 34:861-876

Dewey, M. and L. Barr. 1962. Intercellular connections between smooth muscle cells: The Nexus. Science 137:670

Dewey, M. and L. Barr. 1964. A study of the structure and distribution of the nexus. J. Cell Biol. 23:553-585

Duncan, G. 1969. The site of ion restricting membranes in the toad lens. Exp. Eye Res. 8:406-412

Dunia, I., C. Ghosh, E. Benedetti, A. Zweers and H. Bloemendal. 1974. FEBS Lett. 45:139-144

Epstein, M. and N. Gilula. 1977. A study of communication specificity between cells in culture. J. Cell Biol. 75:769-787

Evans, W. and J. Gurd. 1972. Preparation and properties of nexuses and lipid-enriched vesicles from mouse liver plasma membranes. Biochem. J. 128:691-700

Finbow, M., S. Yancey, R. Johnson and J. Revel. 1980. Independent lines of evidence suggesting a major gap junction protein with a molecular weight of 26,000. Proc. Natl. Acad. Sci. 77:970-974

FitzGerald, P. and D. Goodenough. (in preparation) Development and characterization of differentiating rat lens cultures for MIP expression and communication competence.

FitzGerald, P.G., D. Bok and J. Horwitz. 1983. Immnocytochemical localization of the main intrinsic polypeptide (MIP) in ultrathin frozen sections of rat lens. J. Cell Biol. 97:1491-1499

FitzGerald, P.G., V. Cirillo and D. Goodenough. 1984. Rat lens cell differentiation in culture: A useful model for the study of MIP and intercellular communication. Invest. Ophthal. Vis. Sci. 25:154

FitzGerald, P.G., D. Bok and J. Horwitz. 1984. Distribution of the main intrinsic polypeptide in lens fiber membrane domains. Proc. of the 42nd Ann. Mtg. of the Electron Microscope Society of America, pp. 122-124, San Fransisco Press, S.F., CA

Flagg-Newton, J., I. Simpson and W. Lowenstein. 1979. Permeability of cell to cell membrane channels in mammalian cell junctions. Science 205:404-407

Friend, D. and N. Gilula. 1972. Variations in tight and gap junctions in mammalian tisues. J. Cell Biol. 53:758-776

Friedlander, M. 1980. Immunological approaches to the study of myogenesis and lens fiber junction formation. Current Topics in Developmental Biology 14:321-358

Gilula, N., O. Reeves and A. Steinbach. 1972. Metabolic coupling, ionic coupling, and cell contacts. Nature 235:262-265

Goodenough, D. and J. Revel. 1970. A fine structural analysis of intercellular junctions in the mouse liver. J. Cell Biol. 45:272-290

Goodenough, D. 1979. Lens gap junctions: A structural hypothesis for non-regulated low resistance intercellular pathways. Invest. Ophthal. Vis. Sci. 18:1104-1122

Goodenough, D. and W. Stoeckenius. 1972. The isolation of mouse hepatocyte gap junctions. J. Cell Biol. 54:646-656

Goodenough, D., J. Dick and J. Lyons. 1980. Lens metabolic cooperation: A study of mouse lens transport and permeability visualized with freeze substitution autoradiography and electron microscopy. J. Cell Biol. 86:576-589

Goodenough, D., D. Paul and K. Culbert. 1978. Correlative gap junction structure birth defects: Original article series, XIV:83-97

Gorin, M., S. Yancey, J. Cline, J. Revel and J. Horwitz. 1984. cDNA cloning of MIP26: A new tool for the study of lens structure and development. Invest. Ophthal. Vis. Sci. 25:153

Gros, D., B. Nicholson and J. Revel. 1983. Comparative analysis of the gap junction protein from rat heart and liver: Is there a tissue specificity of gap junction? Cell 35:539-549

Hanna, C. 1965. Changes in RNA, DNA, and protein synthesis in the developing lens. Invest. Ophthal. Vis. Sci. 4:480-491

Hanna, R.B., G.D. Pappas and M. Bennett. 1984. The fine structure of identified electrotonic synapses following increased coupling resistance. Cell Tissue Research 235:243-249

Harding, J. and K. Dilley. 1976. Structural proteins of the mammalian lens: A review with emphasis on changes in development, aging and cataract. Exp. Eye Res. 22:1-73

Henderson, D., H. Eibel and K. Weber. 1979. Structure and biochemistry of mouse hepatocyte gap junctions. J. Mol. Biol. 132:193-218

Hertzberg, E. and N. Gilula. 1979. Isolation and characterization of gap junctions from rat liver. J. Biol. Chem.

Hertzberg, E. and N. Gilula. 1982. Liver gap junctions and lens fiber junctions: Comparative analysis and calmodulin interaction. Cold Spring Harbor Symp. Quant. Biol. 46:639-645

Hertzberg, E., D. Anderson, M. Friedlander and N. Gilula. 1982. Comparative analysis of the major polypeptides from liver gap junctions and lens fiber junctions. J. Cell Biol. 92:53-59

Hertzberg, E. 1980. Biochemical and immunological approaches to the study of gap junctional communication. In Vitro 121:1057-1067

Hyde, A., B. Blondel, A. Matter, J. Cheneval, B. Filloux and L. Girardier. 1969. Homo- and heterocellular junctions in cell cultures: An electrophysiological and morphological study. Prog. Brain Res. 31:283-311

Janssen-Timmen, U., R. Dermeitzel, U. Frixen, A. Leibstein. O. Traub and K. Willecke. 1983. Immunocytochemical localization of the gap junction 26k protein in mouse liver plasma membranes. EMBO Journal 2:295-302

Kensler, R. and D. Goodenough. 1980. Isolation of mouse myocardial gap junctions. J. Cell Biol. 86:755-764

Kinsey, V. and D. Reddy. 1965. Studies on the crystalline lens Invest. Ophthal. Vis. Sci. 4:104-116

Kistler, J. and S. Bullivant. 1980. Lens gap junctions and orthagonal arrays are unrelated. FEBS Letters 111:73-78

Kuszak, J., J. Alcala and H. Maisel. 1981. Biochemical and structural features of chick lens gap junctions. Exp. Eye Res. 33:157-166

Kuszak, J., H. Maisel and C. Harding. 1978. Gap junctions of chick lens fiber cells. Exp. Eye Res. 27:495-498

Kuszak, J., J. Rae, B. Pauli and R. Weinstein. 1982. Rotary replication of the lens gap junction. J. Ultrastruct. Res. 81:249-256

Kuwabara, T. 1975. The maturation of the lens cell: A morphologic study. Exp. Eye Res. 20:427-443

Lasser, A. and E. Balzs. 1972. Biochemical and fine structure studies on the water insoluble components of the calf lens. Exp. Eye Res. 13:292-308

Leeson, T. 1971. Lens of the rat eye: An electron microscope and freeze-etch study. Exp. Eye Res. 11:78-82

Lo, C. and N. Gilula. 1979. Gap junctional communication in the post implantation mouse embryo. Cell 18:411-422

Lowenstein, W. 1979. Junctional intercellular communication and the control of growth. Biochim et Biophys Acta 560:1-65

Lowenstein, W. 1966. Permeability of membrane junctions. Annals, New York Academy of Science 441-472

Lowenstein, W. and Y. Kanno. 1964. Studies on an epithelial gland cell junction. J. Cell Biol. 22:565-586

McAvoy, J. 1980. Induction of the eye lens. Differentiation 17:137-149

Michalke, W. and W. Lowenstein. 1971. Communication between cells of different types. Nature 232:121-122

Miller, T. and D. Goodenough. 1985. Gap junctional structure after experimental alterations in junctional channel conductance. J. Cell Biol. (in press)

Mousa, G. and J. Trevethick. 1977. Differentiation of T4 Rat Lens epithelial cells in tissue culture. Developmental Biology 60:14-25

Nicholson, B., L. Takemoto, M. Hunkapiller, L. Hood and J. Revel 1983. Differences between liver gap junction protein and the lens MIP 26 from rat: Implications for tissue specificity of gap junctions. Cell 32:967-987

Nikaido, H. and E. Rosenberg. 1984. Functional reconstitution of lens junction proteins into proteoliposomes. Proc. of 42nd Ann. Mtg. of the Electron Microscope Society of America, pp. 130-132

Okada, T., G. Eguchi and M. Takeichi. 1973. The retention of differentiated properties by lens epithelial cells in clonal cell culture. Dev. Biol. 34:321-333

Paul, D. and D. Goodenough. 1983. Preparation, characterization and localization of an antisera against bovine MP26, an integral membrane protein from lens fiber cell plasma membranes. J. Cell Biol. 96:625-631

Payton, B., M. Bennett and G. Pappas. 1969. Permeability and structure of junctional membranes at an electrotonic synapse. Science 166:1641

Peracchia, C. and L. Peracchia 1980. Gap junction dynamics: reversible effects of hydrogen ions. J. Cell Biol. 87:719-727

Peracchia, C. 1977. Gap junctions: Structural changes after uncoupling procedures. J. Cell Biol. 72:626-641

Peracchia, C. and L. Peracchia. 1980. Gap junction dynamics: Reversible effects of divalent cations. J. Cell Biol. 87:708-718

Peracchia, C. and S. Girsch. 1984. Calmodulin-mediated gating of lens gap junction channels in vesicles. Proc. of 42nd Ann. Mtg. of the Electron Microscope Society of America, pp 134-137

Peracchia, C. 1978. Calcium effects on gap junction structure and cell coupling. Nature 271:669-671

Phillipson, B., L. Hanninen and E. Balazs. 1975. Cell contacts in human and bovine lenses. Exp. Eye Res. 21:205-219

Piatigorski, J., H. Webster and S. Craig. 1972. Protein synthesis and ultrastructure during the formation of embryonic chick lens fibers In Vitro. Develop. Biol. 27:176-189

Piatigorski, J. 1981. Lens differentiation in vertebrates. A review of cellular and molecular features. Differentiation 19:134-153

Pitts, J. and J. Simms. 1977. Permeability of junctions between animal cells. Exp. Cell Res. 104:153-163

Rae, J. 1973. The potential difference of the frog lens. Exp Eye Res. 15:485-494

Rae, J. 1974. The movement of procion dyes in the crystalline lens Invest. Ophthal. Vis. Sci. 13:147-150
Rae, J. 1979. The electrophysiology of the crystalline lens. Current Topics in Eye Research 1:37
Rae, J. and J. Kuszak. 1983. The electrical coupling of epithelium and fibers in the frog lens. Exp. Eye Res. 36:317-326
Rae, J., R. Thomson and R. Eisenberg. 1982. The effect of 2-4 dinitrophenol on cell-cell communication in the frog lens. Exp. Eye Res. 35:597-609
Rafferty, N. and E. Esson. 1974. An electron microscope study of adult mouse lens: Some ultrastructural specializations. J. Ultrastructural Res. 46:239-253
Raviola, E., D. Goodenough and G. Raviola. 1980. Structure of rapidly frozen gap junctions. J. Cell Biol. 87:273-279
Revel, J. and M. Karnovsky. 1967. Hexagonal arrays of subunits in intercellular junctions of the mouse heart and liver. J. Cell Biol. 33:7-12
Rose, B. and W. Lowenstein. 1975. Permeability of cell junctions depends on local cytoplasmic calcium activity. Nature 254:250-252
Rose, B. and R. Rick. 1978. Intracellular pH, intracellular free Ca, and junctional cell-cell coupling. J. Membrane Biology 44:377-415
Rose, B., I. Simpson and W. Lowenstein. 1977. Calcium ions produce graded changes in permeability of membrane channels in cell junctions. Nature 267:625-628
Russel, P., W. Robison and J. Kinoshita. 1981. A new method for the rapid isolation of the intrinsic membrane proteins from the lens. Exp. Eye Res. 32:511-516
Schuetze, S. and D. Goodenough. 1982. Dye transfer between cells of the embryonic chick lens becomes less sensitive to CO_2 treatment with development. J. Cell Biol. 92:694-705
Schwarzzmann, G., H. Wiegandt, B. Rose, A. Zimmerman, D. Ben-Haim and W. Lowenstein. 1981. Diameter of cell-cell junctional channels as probed with neutral molecules. Science 213:551-554
Simpson, I., B. Rose and W. Lowenstein. 1977. Size limit of molecules permeating the junctional membrane channels. Science 195:294-296
Smelser. G. 1965. Embryology and morphology of the lens. Invest. Ophthal. Vis. Sci. 4:398-410
Spray, D., A. Harris and M. Bennet. 1981. Gap Junctional conductance is a simple and sensitive function of intracellular pH. Science 211:712-715
Subak-Sharpe, H., R. Burk and J. Pitts. 1969. Metabolic cooperation between biochemically marked mammalian cells in tissue. J. Cell Sci. 4:353-367
Takemoto, L. and J. Hansen. 1981. Gap junctions from the lens: Purification and characterization by chemical cross-linking reagent. Biochem. Biophys. Res. Commun. 99:324-331
Takemoto, L., J. Hansen and J. Horwitz. 1981. Interspecies conservation of the MIP of the lens membrane. Comp. Biochem. Physiol. 68B:101-106
Traub, O. and K. Willecke. 1982. Cross reaction of antibodies against liver gap junction protein with lens fiber junction protein with lens fiber junction protein (MIP) suggests structural homology between these tissue specific gene products. Biochem. Biophys. Res. Commun. 109:895-901

Tsunematsu, Y., H. Fukui and J. Kinoshita. 1978. Studies on primary cultures of adult lens cells from normal and hereditary cataractous mice. Exp. Eye Res. 26:671-685

Turin, L. and A. Warner. 1980. Intracellular pH in early xenopus embryos: Iits effect on current flow between blastomeres. J. Physiol 300:489-504

Turner, L. and A. Warner. 1977. Carbon dioxide reversibly abolishes ionic communication between cells of early amphibian embryo. Nature 270:56-57

Unakar, N. and J. Tsui. 1980. Sodium potassium dependent ATPase I. cytochemical localization in normal and cataractous lenses Invest. Ophthal. Vis. Sci. 19:630

Unwin, P. and G. Zampighi. 1980. Structure of the junctions between communicating cells. Nature 283:545-549

Waggoner, P. and H. Maisel. 1978. Immunofluorescent study of a chick lens fiber cell membrane polypeptide. Exp. Eye Res. 27:151-157

Wanko, T. and M. Gavin. 1959. Electron microscope study of the lens fibers. J. Biophys. and Biochem. Cytol. 6:97-102

Warner, A. and P. Lawrence. 1982. Permeability of gap junctions at the segmental border in insect epidermis. Cell 28:243-252

Weir, M. and C. Lo. 1982. Gap junctional communication compartments in the drosophila wing disk. Proc. Natl. Acad. Sci. 79:3232-3235

Young, R.R. and H.W. Fulhorst. 1966. Regional differences in protein synthesis within the lens of the rat. Invest. Ophthal. Vis. Sci. 5:288-297

Zampighi, G., S. Simon, J. Robertson, T. McIntosh and M. Costello. 1982. On the structural organization of the isolated bovine lens fiber junction. J. Cell Biol. 93:175-189

Zampighi, G. and J. Robertson. 1973. Fine structure of the synaptic discs separated from the goldfish medulla oblongata. J. Cell Biol. 56:92-105

Zampighi, G. and P. Unwin. 1979. Two forms of isolated gap junctions. J. Mol. Biol. 135:451-464

Building an Ommatidium: Geometry and Genes

Donald F. Ready, Andrew Tomlinson, and Richard M. Lebovitz

The intricate network of the nervous system is woven layer upon layer during development; each stage of assembly serves as the scaffold upon which the next stage built. Many cellular and molecular mechanisms cooperate in this process, but unravelling their contributions is often hindered by the heterogeneity of neural tissue and the complexity of the neural network. The unit eye, or ommatidium, of the Drosophila compound eye is a simple network of interconnected cells that offers several advantages for studying these questions. Each ommatidium is a stereotyped assembly of a small number of photoreceptors and accessory cells in which each cell is precisely positioned by the contacts it makes with its neighbors; the remarkable regularity of the fly compound eye is the manifestation of a precise underlying pattern of cellular connectedness. The several hundred-fold reiteration of the ommatidium makes the compound eye a sensitive amplifier of this network. A wealth of genetic variants that perturb normal fly eye development has been isolated making these genes amenable to a host of genetic and molecular techniques. Finally, ommatidial assembly occurs in an easily accessible retinal epithelium and has been well characterized morphologically. We are attempting to exploit these advantages to understand how flies build the network of selective associations that shape an ommatidium.

Ommatidia assemble in an integrated program that operates across cellular and molecular levels, and an understanding of eye development will require a coordinated approach using methods appropriate to both dimensions. We have have applied a combination of anatomical, monoclonal antibody and genetic techniques to trace the outlines of ommatidial assembly. The precise, highly amplified anatomy of the ommatidium makes it an excellent target for the convergence of these diverse approaches since results from each technique can be cross-correlated with others on the regular spatial and temporal map of the developing eye.

Our picture of ommatidial assembly is far from complete, and much additional work will be required to understand the logic of fly eye development. This endeavor seems well worth the effort. The developmental processes of ommatidial assembly in Drosophila are typical of those seen in more complex neural epithelia, including the human nervous system. An understanding of pattern formation in the fly eye may be of value in understanding network-building in these systems and the congenital abnormalities that occur when this building goes awry.

ANATOMY OF THE COMPOUND EYE

The quality of the image obtainable by a compound eye depends strongly on the regularity of its ommatidial array, and the eyes of highly visual flying insects have been honed to almost crystalline precision. Externally, the Drosophila compound eye presents a smooth array of approximately 750 hexagonal facets (Fig. 1). The array is organized along two oblique axes and one horizontal axis. Each facet is a corneal lens that images a small portion of the visual world onto eight underlying photoreceptors (R1-R8) and a panoramic view is reconstructed in the first optic brain, or lamina by the superposition of overlapping ommatidial images (Trujillo-Cenoz and Melamed (1966); Braitenberg, 1967).

The fly eye also carries a regular field of small mechanosensory bristles on its surface. The sensory neurons of these bristles project to a distinct, non-optic lobe of the brain; touching an eye bristle causes the fly to clean the eye with a foreleg. The cells forming the mechanosensory bristles appear to be developmentally equivalent to a type of pigment cell, the tertiary pigment cell (see below), and in the following discussion the two are treated interchangeably.

A Drosophila ommatidium is a roughly cylindrical assembly of concentrically arranged cells (Figs. 2,3). At the core of each ommatidium lie two "inner" receptors, R7 and R8, surrounded by a ring of six "outer" receptors, R1-R6. The axial face of each photoreceptor is elaborated into a stack of photosensitive,

Fig. 1. Scanning electron micrograph of a Drosophila compound eye. Left eye, anterior to the right. Calibration bar equals 10 microns.

rhodopsin-bearing microvilli, the rhabdomere. Rhabdomeres of the six outer receptors extend the depth of the retina. Rhabdomeres of R7 and R8 lie along the central axis of the ommatidium, R7 above R8. Genetic and physiological evidence indicates that the outer

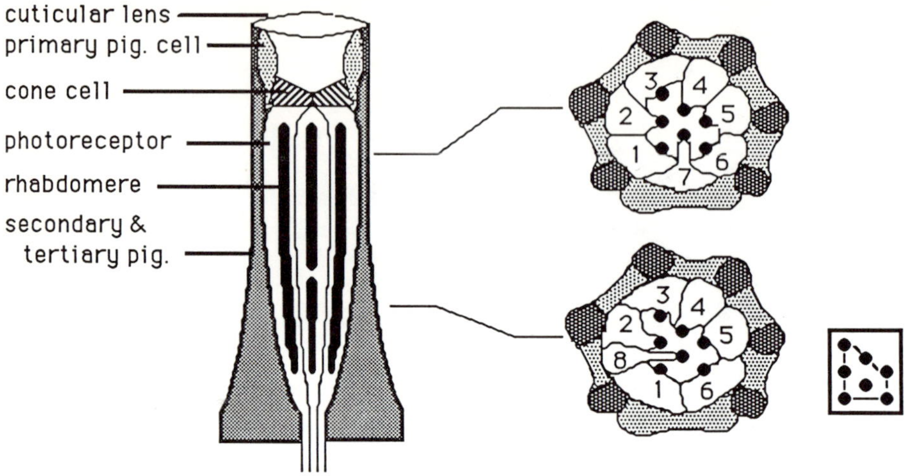

Fig. 2. Schematic ommatidium. Inset shows asymmetric, trapezoidal arrangement of rhabdomeres.

receptors form a group distinct from the inner receptors, which are also distinct from each other (Harris et al., 1976).

Rhabdomeres of the outer receptors are arranged in an asymmetric, trapezoidal pattern that gives each receptor a unique positional identity (Dietrich, 1909). Trapezoids in dorsal and ventral ommatidia are arranged in mirror-symmetrical forms that meet along an equator that divides the eye roughly in half. The functional significance of this pattern inversion is not known, but it provides a useful reference for describing the positions of ommatidial cells. Thus, R1, R7, and R6 lie on the equatorial side of each ommatidium, while R3 and R4 lie on the polar side, facing either the dorsal or ventral pole of the eye.

Four cone cells lie above the photoreceptors. These cells secrete the corneal lens and a refractile body, the cone, that together form the dioptric system of the ommatidium. Although these cells are positioned above the photoreceptors in the adult ommatidium, earlier in development they encircle the photoreceptors.

Two primary pigment cells form the next ring of cells, forming a collar around the cone cells.

The final ring of cells is formed by secondary and tertiary pigment cells. Secondary pigment cells are shared between two ommatidia and form the sides of the hexagonal chambers which

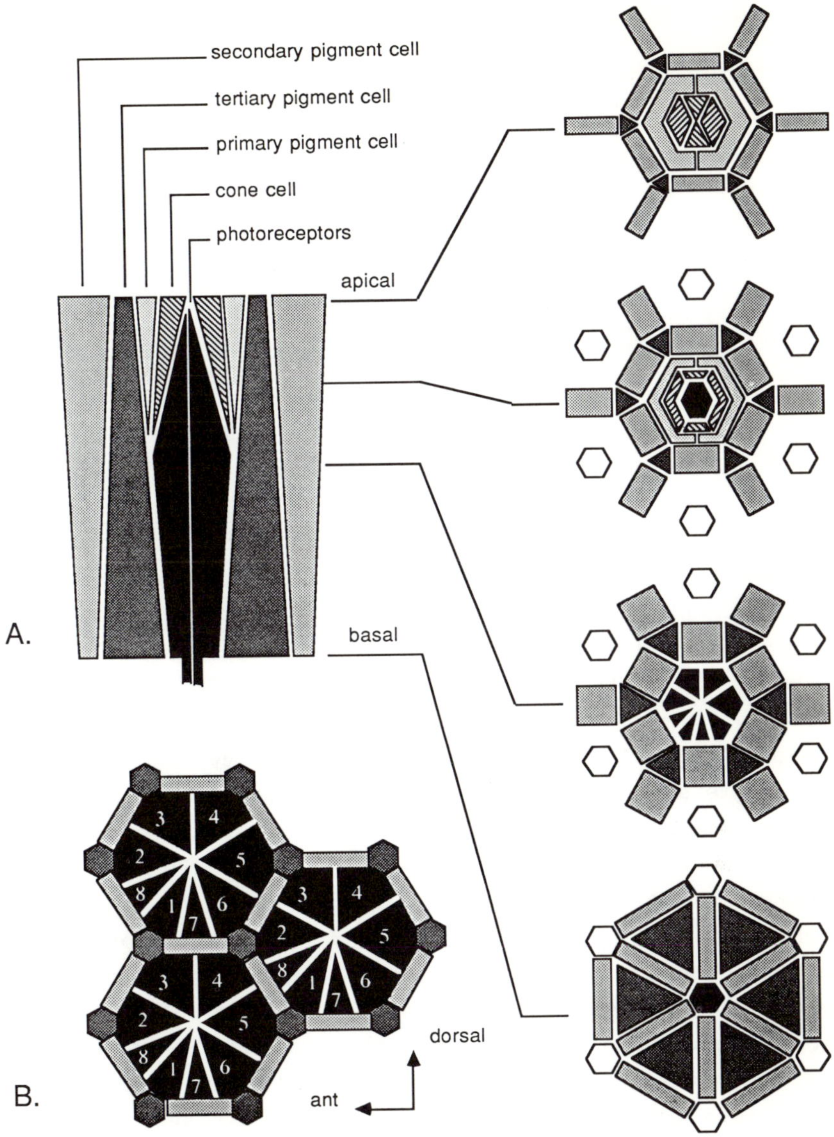

Fig. 3. Schematic ommatidium showing cell contacts. A. Side and cross-sectional views. B. Detail to show photoreceptor positions in ommatidium.

surround each ommatidium. Tertiary pigment cells lie at facet vertices, and are shared among three neighboring ommatidia. The sheath of pigment cells surrounding each photoreceptor core enhances an ommatidium's restricted view by insulating it from off-axis illumination.

Each Drosophila ommatidium is thus an assembly of twenty-two uniquely identifiable cells. The assignment to a single ommatidium of three secondary pigment cells, one tertiary pigment cell (and one mechanosensory bristle group) is an accounting convenience; these shared cells form a seamless honeycomb of chambers in which the unshared primary pigment cells, cone cells and photoreceptors are embedded.

CELL-CELL CONTACTS

The precise position a cell occupies within an ommatidium is an expression of the stereotyped contacts it makes with its neighbors; the ommatidial network is the simultaneous "solution" of contacts made among a small population of cells, each with a stereotyped set of partners. The regularity of the cellular mosaic of the Drosophila retina was noted by Waddington (1962). Fig. 4 is a drawing of a tangential section of a late pupal eye based on an electron micrograph by Margaret Perry (Waddington, 1962, plate xxi). The plane of this section is in the outer portion of the eye, where the rhabdomere of R7 occupies the ommatidial center. Secondary pigment cells are lightly shaded and tertiary and hair-nerve group cells are more heavily shaded. At this level, axons of the hair-nerve groups are very small in cross section and are omitted from the drawing at several locations.

The regularity of cell contacts in the eye is evident on inspection. For example, every horizontal secondary pigment cell contacts R1, R7 and R6 along its polar face and R3 and R4 on its equatorial face. Secondary pigment cells forming the oblique sides of the ommatidial hexagon show a similar stereotypy of contacts. Tertiary pigment cells are distinguished by a ring of contacts; around their perimeters R2, R4 and R6 in three neighboring ommatidia alternate with three secondary pigment cells lying along the three axes of the eye. Deviations from the standard pattern are also evident. In the plane of this section, R1 occasionally occludes the contact of R2 with its usual secondary pigment cell partner. We are currently investigating the nature of this "noise" in retinal cell contacts.

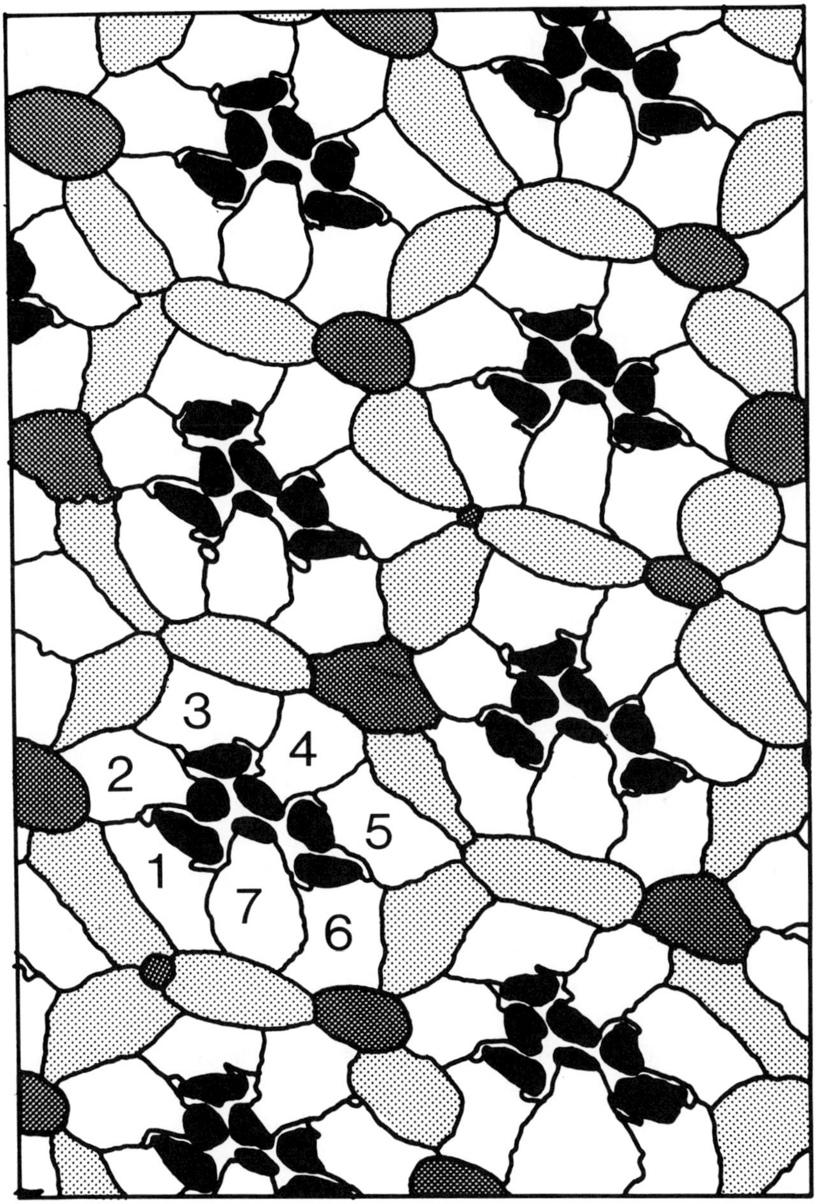

Fig. 4. Tangential section through eye. Redrawn from an electron micrograph by Margaret Perry (Waddington, 1962, plate xxi). Photoreceptors are numbered. Rhabdomeres black; secondary pigment cells light gray; tertiary pigment cells and hair nerve groups dark gray.

To map cell contacts more closely, we have reconstructed ommatidia using serial thin sections of pupal eyes. (Pupal eyes are preferable to adult eyes for this purpose since the pigment granules which accumulate late in development contort cell membranes, making contacts difficult to reconstruct.) A schematic (and slightly simplified) ommatidium based on these reconstructions is shown in Fig. 3. Fig. 3B shows an ommatidial cross-section equivalent to that in Fig 4, with the exception that R8 is also indicated. An ommatidial side view, with four cross-sections, is shown in Fig. 3A.

Contacts between retinal cells appear quite tenacious, and are maintained throughout the depth of the eye, despite changing cell shapes. Perhaps the most notable change in cell shape is the constriction of the photoreceptors toward the base of the eye as they neck down to form axons. Their contribution to an ommatidial cross section changes from the dominant profile at the middle of the eye, to a small, centrally located axon bundle at the base. These changes are matched by changes in surrounding pigment cells. For example, a tertiary pigment cell contacting R2, R4, and R6 is drawn into a widening triangular profile toward the inner surface of the retina. Secondary pigment cells also maintain their photoreceptor contacts throughout the depth of the retina. As a consequence, the long axis of their cross-sectional profiles are transformed by ninety degrees between the outer and the inner surface of the eye. For example, the horizontally-oriented secondary pigment cell that contacts R1, R7, and R6 on one side and R3 and R4 on the other is drawn into a vertical orientation at the base of the eye.

Polyhedra provide a convenient representation of the contacts made by ommatidial cells (Fig. 5). Each polyhedral face is an idealized contact between neighboring cells. (In order to keep the drawing manageable, certain contact faces of relatively minor area have been omitted. For example, the tertiary pigment cell would be more accurately represented by a six-sided pyramid, allowing for its contacts to R2 ,R4, and R6.) Although an abstraction, cell shapes in this model are sufficiently close to their living counterparts to be recognizable. When assembled, the shapes form a "cubist" ommatidium.

In this representation, the honeycomb of shared pigment cells that forms the continuous mesh of the compound eye is formed by an alternating repetition of two shapes: secondary cell tetrahedrons and tertiary cell pyramids. At each facet vertex, three faces of a

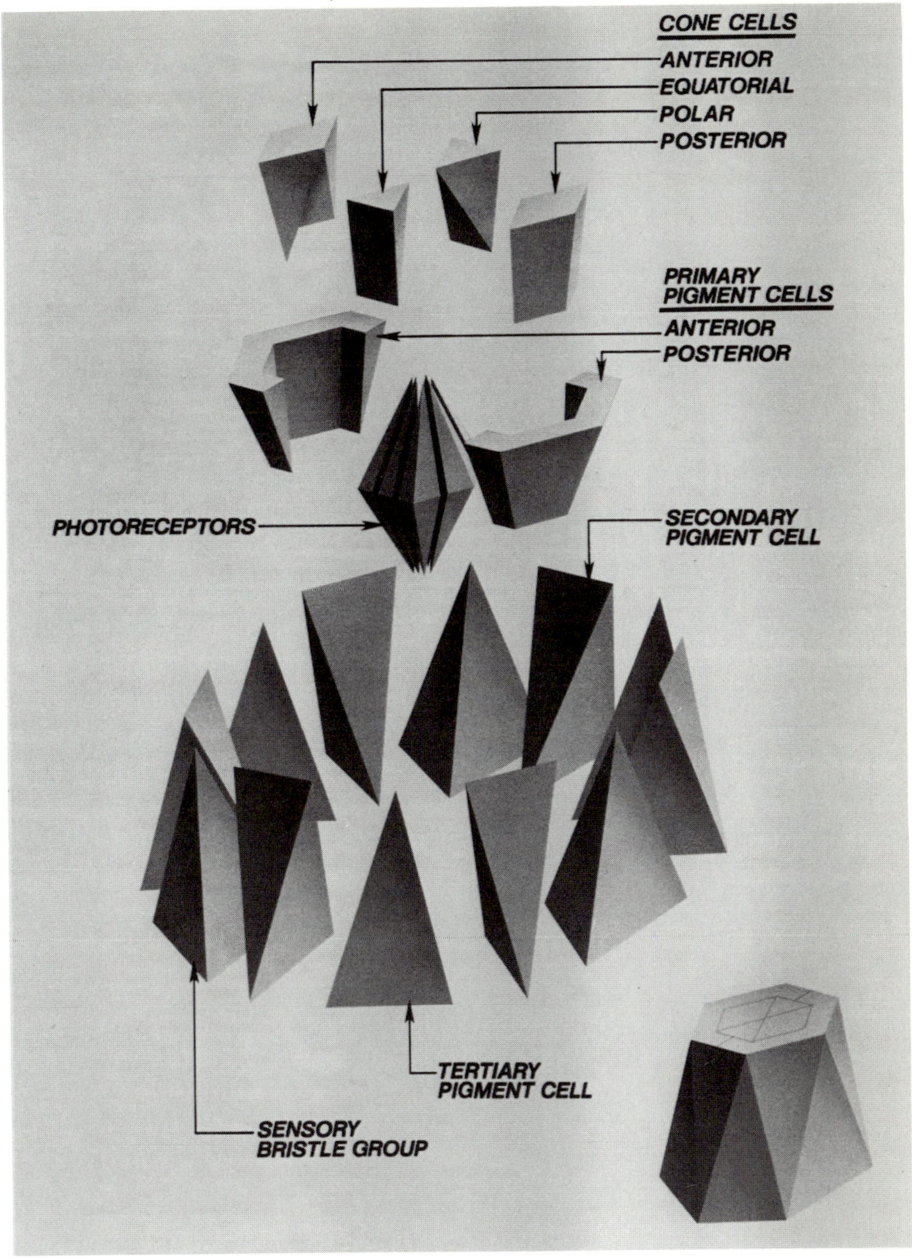

Fig. 5. Exploded view of ommatidium.

tertiary cell are apposed by faces of three secondary cells. The internal wall of an ommatidial chamber is formed by faces of six secondary pigment cells, and the "edges" of three tertiaries and three hair-nerve groups.

Photoreceptors taper at their outer ends as well their axonal ends. At the outer surface of the eye two primary pigment cells and four cone cells are interposed between the photoreceptors and the secondary and tertiary cells. The anterior and posterior primary pigment cells display equivalent contacts and each can be represented by a single shape that forms one half of the collar between the secondary pigment cells and the cone cells. The anterior and posterior cone cells are also of only a single polyhedron type. Each is cradled by a single primary pigment cell along its centrifugal surface and contacts either receptors R1, R2, and R3 or R4, R5, and R6 on its axial face. The polar and equatorial cone cell shapes are likewise equivalent and both can be represented by a single shape. The centrifugal faces of the equatorial and polar cone cells contact both primary pigment cells and on their axial faces contact receptors R1, R7, and R6, or R3 and R4, respectively.

The final elements in this schema are wedge-shaped photoreceptors which fill the internal chamber of an ommatidium like the sections of an orange. Each of the outer six receptors contacts two neighboring receptors along its flanks and is backed into a corner of the internal chamber, contacting two secondary pigment cells and a tertiary (or a mechanosensory bristle group) along its centrifugal surface.

DEVELOPMENT OF THE COMPOUND EYE

The Drosophila compound eye, and the fly that carries it, require roughly nine days of development to complete; one day of embryogenesis, four days of larval life, and four days of pupal life intervene between the fertilization of a fly egg and the emergence of an adult. Mid-way through embryogenesis, a plate of about 20 epithelial cells, the eye disc, is allocated to produce the retina. Over the next three days, cell proliferation expands the disc to an unpatterned epithelium of approximately 2000 cells (Becker, 1957). Beginning at the posterior of the disc, a wave of morphogenesis then advances across the eye disc and transforms it into a highly organized field of ommatidial precursors. Approximately two days are required to generate the adult number of ommatidial precursors. The remainder of development, the majority of pupal life, is spent in the maturation of precursors into adult ommatidia. During this time, cells elaborate characteristic retinal structures such as rhabdomeres, lenses and pigment.

Two explanations have been advanced to account for the orderly, anterior growth of insect compound eyes, including Drosophila's. One model proposes that a proliferating band of stem cells along the growing margin of the eye spins out progeny committed to mature into retina (Bodenstein, 1953; Nowel and Shelton 1980). The alternative proposes that the growing edge of the eye has an instructive role, inducing or recruiting cells to join the developing eye (Wolsky, 1949; White, 1961; Hyde, 1972; Ready et al., 1976; Lawrence and Shelton, 1975). In the first case, the retina grows by accretion as cells are "pushed" into place from a specialized anterior margin of "retinoblasts"; in the second, the leading edge of the pattern "pulls" uncommitted cells into the retina.

The growing edge of the organized eye pattern in Drosophila is clearly visible as a distinct front, marked by an indentation or furrow running across the prospective retinal epithelium (Figs. 6,10). Ahead of the furrow, the tissue is unpatterned, while behind the furrow the epithelium is organized into ommatidial precursors. The abrupt emergence of the eye pattern along this front, together with the observation that a cell's fate in the developing eye depends upon its position, rather than its ancestry, prompted the suggestion that the fruit fly eye develops like a crystal (Ready et al., 1976). The leading edge of the array was postulated to act as a template against which unpatterned cells were organized and thereby informed of their fate. Cells newly incorporated into the lattice in turn informed cells ahead of them of their proper roles, propagating the pattern forward via a chain of inductions.

In order to test this idea, we broke the postulated inductive chain by transplanting fragments of eye discs from the region ahead of the furrow into host larvae. After host larvae pupated and emerged as adults, the implants were recovered and examined for the presence of ommatidia. If the crystal-growth hypothesis is correct, these fragments should not form ommatidia. The results were unambiguous; the majority of fragments taken from ahead of the furrow and separated from the leading edge differentiated ommatidia when cultured in host larvae. The crystal-like regularity of the Drosophila compound eye thus does not arise like a crystal.

Morphogenesis in the eye proceeds according to a tight

timetable in which posterior cells are developmentally more advanced than anterior ones. Progressively more mature precursors are located at increasing distances behind the furrow. As a result, the temporal sequence of ommatidial development is displayed spatially. In order to examine this temporal map in cultured disc fragments, implants were recovered from hosts before they pupated and ommatidial precursors were visualized using a monoclonal antibody that stains photoreceptors (Lebovitz and Ready, 1982). In these fragments, ommatidial formation progressed as it would have in situ; posterior regions always differentiated before more anterior areas. The stability of the course taken by the wave of differentiation in these fragments suggest a regular schedule of differentiation is mapped onto the prospective retinal epithelium ahead of the furrow. A program of retinal differentiation appears to be under way well before overt ommatidial precursors are assembled at the furrow.

The transplant results obtained in developing Drosophila eyes are compatible with a meristematic eye margin described in the cockroach by Nowel and Shelton (1980) and Shelton et al. (1983). In the cockroach, new ommatidia are added to the eye from a specialized margin at the growing edge. A mitotically active "proliferation zone" at the growing margin produces cells which then assemble into ommatidial clusters in a "maturation zone" and develop further to form adult ommatidia. The youngest ommatidia lie anterior and more mature ommatidia to the posterior. If the pre-furrow region of the Drosophila eye disc is a comparable proliferation zone, disc fragments including this territory should be able to differentiate ommatidia, as observed.

The ability of pre-furrow eye disc fragments to differentiate ommatidia independent of contact with the advancing pattern appears at first to be in conflict with White's (1961) observation that head epidermis from a region destined to be occupied by the developing mosquito compound eye is unable to differentiate ommatidia when cultured in a host. In order for implants to form retina, they had to include a portion of the eye primordium, the optic placode. A comparison of histological sections of eye discs with that shown by White suggests the leading edge of mosquito optic placode corresponds to the pre-furrow eye territory of the disc. If the mosquito eye expands into territory occupied by unspecialized head epidermis by mitotic activity of its anterior margin rather than an inductive recruitments, implants without fragments of the placode would fail to form ommatidia because they lacked retinoblasts.

Drosophila eye pattern defects caused either by mutations or environmental treatments, also argue against a template-like activity of the growing edge. Disturbances of the lattice rarely propagate significant distances "downstream." Becker (1957) noted that X-irradiation of early third instar larvae could produce a vertical band of ommatidial disruption, or roughness which was restricted to the posterior of the eye. A similarly restricted roughening of the eye was also observed following cold shock in a temperature-sensitive mutant, Notch (Foster and Suzuki, 1970). Heat-pulses applied to the temperature-sensitive mutant shibire can produce a dorso-ventral scar, ahead of which lies a normal eye pattern (Poodry et al., 1973; Fristrom and Fristrom, 1982). Mosaic eyes containing patches of mutant tissue in a normal background display roughness restricted primarily to the mutant area, with little effect on adjoining retina of normal genotype (Meyerowitz and Kankel, 1978). The observation that single ommatidia possessing approximately normal morphology can be found separated from other ommatidia in severely affected eye mutants such as $Drop^{Micropthalmia}$, also indicates that ommatidia do not require adjoining patterned tissue for assembly.

POSITIONAL DETERMINATION OF RETINAL CELL TYPE

Although the transplant experiments demonstrate that the eye does not grow via a template activity of the leading edge of the pattern, genetic marking experiments show the fate of a retinal cell is determined by positional cues rather than its ancestry (Ready, et al., 1976). The descendents of an individual cell in the developing retinal epithelium can be readily identified with genetic markers and the contribution of this clone to the adult eye can be scored cell by cell. Such studies reveal no fixed pattern of lineage; in chimeric eyes, ommatidia are built from both marked and unmarked clones of cells in apparently random mixtures. Even clones marked late in development, while the retinal pattern is spreading across the eye disc, show no separation into lineages destined to produce a particular cell type (Lawrence and Green, 1979). Cell marking experiments also show that clones of retinal cells remain together to form coherent patches, rather than mixing in salt and pepper fashion with neighboring cells. Taken together these results suggest cells of the retinal epithelium determine their appropriate cell type in place based on local interactions. As discussed below, the stereotyped sequence and morphology of ommatidial assembly may endow local interactions with ample information for this task.

THE MORPHOLOGICAL PICTURE OF ASSEMBLY

Along the morphogenetic front, ommatidial precursors first appear as stereotyped aggregations or condensations of five cells embedded in the surrounding epithelium. These preclusters each contain cells destined to produce receptors R2,R3,R4,R5, and R8. From the outset, ommatidial cells make accurate, reproducible contacts. R8 is centrally placed, with R2 and R5 making large, flanking contacts. R3 and R4 make smaller contacts with R8 and form long, flanking contacts with R2 and R5 respectively. In the next stage of ommatidial assembly photoreceptors R1, R6, and R7 are added to form an eight-cell cluster. As these cells join the cluster, they occupy well defined "niches" formed by cells already incorporated into the precursor. For example, as the cell destined to form R7 is added to the photoreceptor cluster, it occupies a unique channel whose walls are formed by R1 and R6 and whose floor is formed by R8.

Non-photoreceptor cells surrounding the photoreceptor core are less readily identified since they lack obvious distinguishing morphological features at this stage, but these cells can also make reproducible contacts with cells of the photoreceptor core. For example, the cell destined to form the anterior cone cell occupies a unique site on the anterior face of the eight-cell cluster, contacting photoreceptors R1, R2, and R3.

Once formed, ommatidial precursors mature in a program of cell movements that permit additional contacts to be formed. For example, the four cone cells are initially incorporated into the ommatidium as a ring of cells surrounding the photoreceptor core; equatorial and polar cone cells do not touch at this stage. As the photoreceptor core subsequently sinks into the epithelium, the ring of cone cells closes, introducing the four cone cells to each other across the ommatidial axis.

The orderly sequence of ommatidial assembly, together with the lack of cell migration revealed in the genetic marking experiments suggest that the fates of undetermined cells in the retinal epithelium are determined in place by their interactions with neighboring cells that have already been incorporated into a precursor. At each stage of assembly, the stereotyped morphology of the precursor presents defined niches to surrounding epithelial cells which could supply the cues necessary for a cell to determine its appropriate fate. For example, a cell in contact with R1,R6, and R8 could be triggered to form R7, while a cell in contact with R1, R2, and R3 could be informed to become an anterior cone cell. Some form of cell identity that allows cells to discriminate among neighbors is a clear requirement for such an assembly mechanism.

NUCLEAR MOVEMENTS AND CELL CONTACTS

The eye disc is a monolayer epithelium; each cell extends from the outer, future corneal surface to the inner surface of the sheet (Fig. 6). Cells contact neighboring cells throughout this depth. The broadest expanse of a cell's surface occurs at the level of its nucleus; above and below this cell body the cell extends processes which are anchored to the inner and outer surface. Consequently, neighboring cells with their nuclei at similar levels in the epithelium will appose broad surfaces, while neighbors with nuclei positioned at different depths will make relatively smaller contacts between the cell body of one cell and the process of the other.

As ommatidia are formed, a choreographed program of nuclear movements redistributes the areas of contact between cells. Ahead of the furrow, nuclei are distributed uniformly throughout the depth of the epithelium, without apparent pattern. Immediately behind the furrow, the nuclei of the five photoreceptors that comprise the precluster segregate into the outer level of the epithelium, maximizing the contacts they make with each other, and diminishing the contacts they make with the non-photoreceptor cells that remain at deeper levels. As R1, R6 and R7 are added, their nuclei migrate up to join those of the precluster.

The upward nuclear movement of cells joining the ommatidial precursor extends to non-photoreceptors as well. After the eight-cell photoreceptor core is assembled, the four cone cells rise along its flanks. As the cone cells rise, the photoreceptor core sinks into the epithelium, relinquishing its outermost position as the cone cells come to lie over it. The primary, secondary, and tertiary pigment cell nuclei move up later in development. An attractive, but untested, idea is that nuclear movements are driven by the tendency of cells to maximize their contacts with more adhesive neighbors; nuclei are dragged along as cell surface interactions appose cell bodies of adhesive cells. If retinal cells become more adhesive as they mature, the concentric addition of cells to the developing ommatidium may be reflected in the observed pattern of nuclear migrations.

The morphological picture of ommatidial assembly suggests a "reverse fountain" in which cells are added concentrically around a sinking center; cells incorporated earlier are displaced downward as later additions migrate upward over them. According to this model, the first cell incorporated into the ommatidium should become the deepest. The extreme inner location of R8's nucleus in the adult ommatidium and its central position in the developing ommatidium are consistent with the idea that R8 acts as a founder cell.

MUTATIONS AND EYE DEVELOPMENT

Despite the comparative simplicity of the ommatidium, its successful assembly depends on the normal function of a large number of genes. Well over 200 genes have been identified that, when mutant, perturb normal eye development. This number excludes the many genes whose malfunction kills flies early in development, but may normally play a role in ommatidial assembly. The large number of genes essential to normal eye development is perhaps not surprising in view of the interlocking program that must be executed to build an ommatidium; defects in the scaffold at any stage of assembly will ripple through further development. The reiterative nature of the compound eye makes it a sensitive amplifier of defects of cell physiology. For example, a subtle change in membrane fluidity could delay a normally scheduled cell movement, leading to the formation of faulty niches and the consequent miscueing of undetermined cells in over seven hundred ommatidia.

The large number of mutations affecting eye development also points to the difficulty of identifying genes "directing" eye development. Most mutations that disturb the eye pattern also affect other tissues. Most also cause the order to decay in ways that roughen the eye generally and do not cause the eye to default into recognizable subpatterns. Such mutants may nevertheless be informative. For example, flies mutant in the gene <u>echinus</u> show wing and bristle defects as well as rough eyes. Sections of *echinus* eyes reveal supernumerary pigment cells. This observation complements the finding that surplus secondary pigment cells are removed from the eye by cell death during pupal life. The normal *echinus* gene may thus participate in the removal of superfluous cells from the developing eye.

One mutant that may exert its effect close to the logic that determines cell type in the eye is the mutant *sevenless*, which, not surprisingly, lacks photoreceptor R7 (Harris et al., 1976). Our reconstruction of the events leading to the absence of R7 in the mutant indicates that a cell comes to occupy a niche in the developing cluster that is appropriate for the cell destined to produce R7 and executes the early cell movements typical of the prospective R7 (fig. 6). However, when the photoreceptor cluster sinks into the retinal epithelium, the mutant "R7" remains in the outer levels of the epithelium and differentiates as a cone cell instead (Tomlinson and Ready, 1985). The niche appears to provide normal cues, since it is necessary only for the prospective R7 to be genetically defective in order to produce the mutant effect. It is not known whether the mutant cell is dyslexic and misreads its niche, or if it defaults into the path leading to cone cell differentiation. The cell-specific transformation of R7's fate in *sevenless* offers a favorable opportunity to study how cells select their appropriate cell types.

CELL CONTACTS AND SYMMETRIES IN DEVELOPING OMMATIDIA

The regular morphology of ommatidial precursors suggests a pairwise equivalence of the six outer receptors: R1=R6, R2=R5, and R3=R4. This equivalence is manifest in both the characteristic positioning of cells in the ommatidial cluster, and in the size of the contacts they make with R8. The eight-cell ommatidial cluster shows strong bilaterality across the future dorso-ventral axis (Fig. 7). To either side of a midline drawn between R3 and R4, and bisecting R7 and R8, the outer receptors are symmetrically positioned. Indeed, in the absence of a directional cue, it is impossible to distinguish between members of a pair. It is notable that the chirality of the adult ommatidium is based on the choice of one of the two "combined" photoreceptor identities; once a distinction is made, the handedness of the ommatidium is determined.

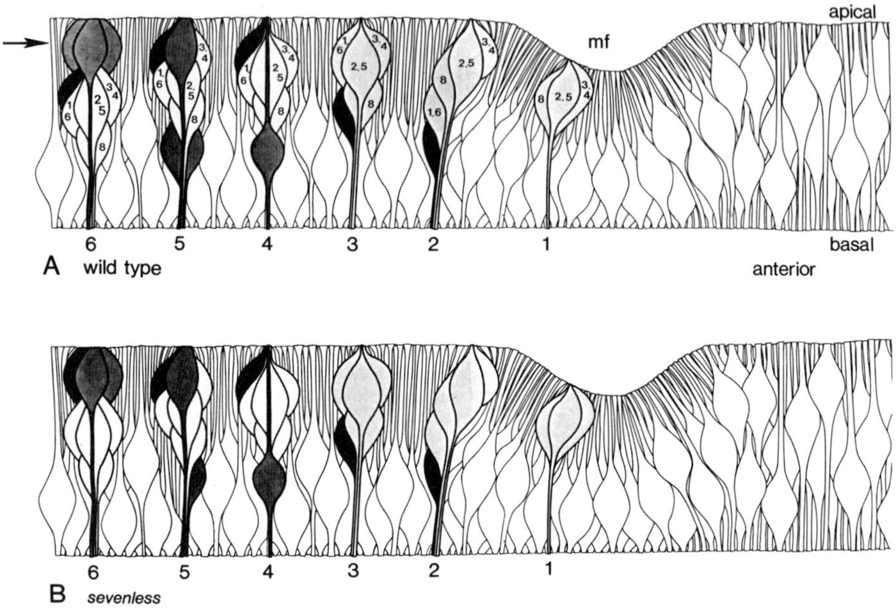

Fig. 6. Ommatidial assembly in wild type (A) and sevenless (B). All cells (except for dividing cells, not shown) extend from the apical surface to the basement membrane of the epithelium. The nuclei are located in the cell bodies. Schematic side view of precursors. Photoreceptors R1-R6 and R8 numbered; Photoreceptor R7: black. Symmetrical pairs R3&R4, R2&R5, and R1&R6 are indicated by single profiles. Cone cells: dark grey; all others: white. mf: morphogenetic furrow. Arrow indicates plane of section shown in Fig. 6. A. 1: five cell precluster. 2: R1,6,7 rise to join the precursor. 3: symmetrical cluster. 4: as R7 completes apical migration two cone cell nuclei rise. 5: two more cone cell nuclei rise. 6: four cone cell nuclei overlie the photoreceptors. B. 1,2,3,4: as above. 5: only a single cone cell nucleus rises. 6: cell in the R7 niche remains apical and joins the cone cell unit, only seven photoreceptors sink basally (From Tomlinson and Ready. Copyright 1985 by the AAAS.)

The areas of the contacts outer receptors make with R8 display a pairwise regularity (Fig. 8). R3 and R4 each make a narrow contact with R8, occupying roughly seven percent of R8's surface. Cells R1 and R6 each cover about fourteen percent of R8's surface and R2 and R5 each claim approximately twenty-one percent. The contact between R8 and R7, is approximately seventeen percent. The differences in the sizes of the contact faces between photoreceptors appears to be related to the identities of the interacting cells,

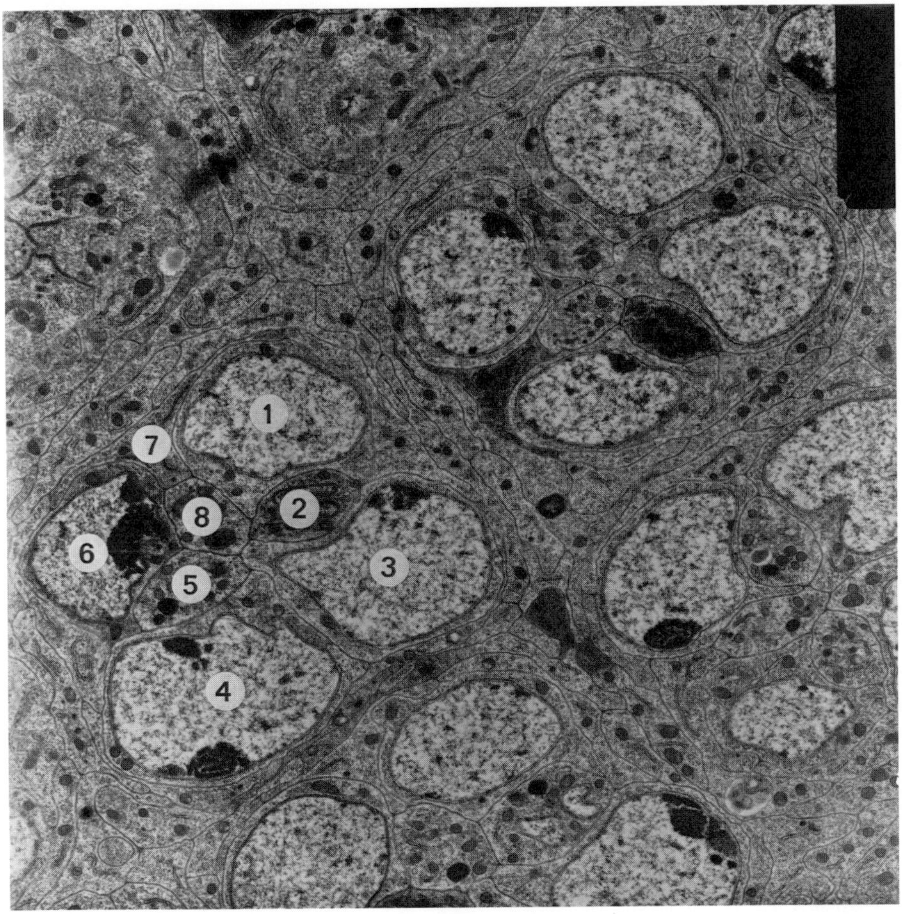

Fig. 7. Electron micrograph showing symmetrical eight-cell clusters (Fig. 5A, 3). Each cluster is bisected by a line passing between R3&R4, R2&R5 and R1&R6, and through the unpaired R7 and R8. In this apical plane, the nuclei of R1,R3,R4 and R6 predominate, nuclei of R2, R5, R7 and R8 are more basally located. In the two clusters shown in toto, R3 occludes the contact of R4 with R8 in this plane. In the two clusters shown partially (right), both R3 and R4 make contact with R8. Note that clusters of both chirality are included in this field. The cluster above and to the right is a mirror image of the numbered cluster. In it, R1 is toward the bottom of the page and cells would be numbered counter-clockwise. Anterior is to the right.

rather than a general "stickiness" or "non-stickiness". For example, R4 simultaneously makes a small contact with R8, and large contacts with its flanking partners, R3 and R5.

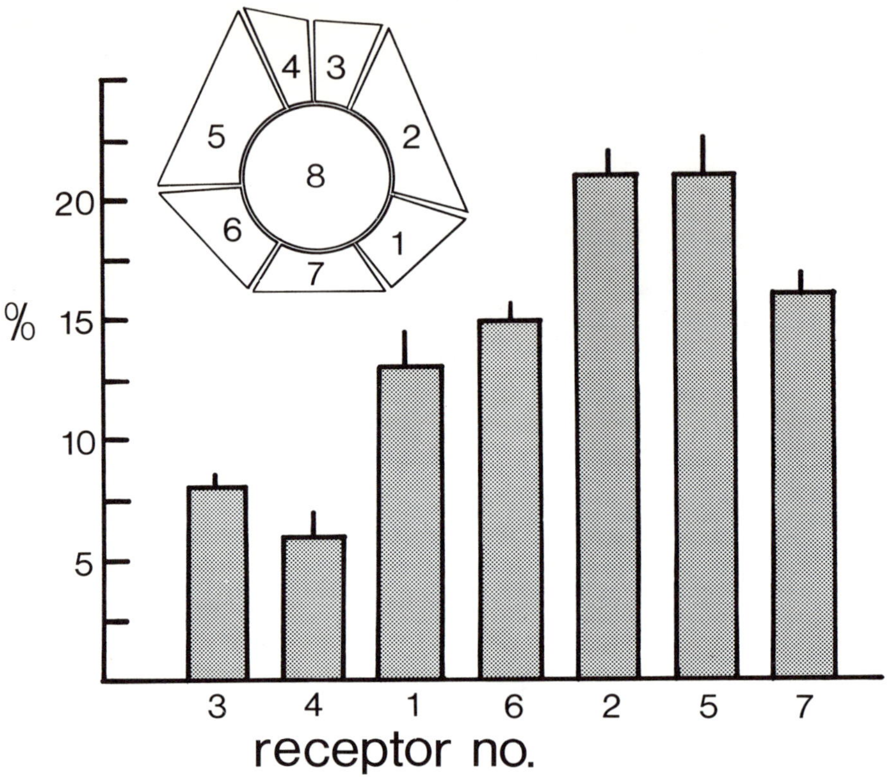

Fig. 8. Contact areas of receptor cells in symmetrical cluster. For each of six neighboring clusters, the contact of each receptor with R8 as well as adjacent receptors was measured from electron micrographs of serial thin sections using a Zeiss MOP III digitizer. The fraction of R8's total surface (in %) occupied by each receptor is shown in the histogram. Contacts of the outer six receptors with R8 fall into three non-overlapping classes, R3&R4, R1&R6, and R2&R5. Error bars indicate standard deviations. In the inset, R8's surface is represented by the circumference of the inner circle, and contacts with surrounding cells are scaled to the appropriate fraction of the circumference. Contacts between adjacent outer receptors are indicated on the same scale. The symmetry of the inset figure derives from the pairwise equality of the contacts outer receptors make with R8.

The initial five-cell pre-clusters containing receptors 2,3,4,5 and 8 are less well characterized, but show the same pairwise positioning of R3 and R4 and R2 and R5 seen in the eight-cell cluster.

The symmetry of the eight-cell cluster is lost as subsequent cell rearrangements build the adult ommatidium. As R8 pushes out anteriorly between R1 and R2 to achieve its location in the adult eye, the pairings of the outer cells are lost and each becomes a uniquely identifiable pattern element. The asymmetry of the adult ommatidium is a broken symmetry.

SYMMETRY IN THE ADULT OMMATIDIUM

With the exception of the trapezoidal rhabdomere pattern, and the eccentric positioning of R8's cell body, the adult ommatidium shows the same bilaterality that characterizes the ommatidial precursor. The decidedly "deliberate" pattern of contacts among retinal cells suggests cells respect underlying commonalities and distinctions. The contacts secondary pigment cells make with their neighbors, as well as genetic and morphological distinctions, can be used to subdivide secondary pigment cells into two categories, "oblique" and "horizontal". Horizontal secondaries contact symmetry partners R1 and R6, or R3 and R4, as well as the equivalent anterior and posterior pigment cells. Oblique secondaries contact non-equivalent photoreceptors, for example R1 and R2, or R4 and R5. Horizontal secondaries can be selectively "collapsed" by the eye mutant *eyeless*^{Russian}, transforming ommatidial hexagons into diamonds formed by four normal oblique secondaries (Ready et al., 1976). A distinction between horizontal and oblique axes is also seen in development. The horizontal axis of the eye lattice is established later in development than the two oblique axes, and the oblique sides of ommatidial hexagons are slightly longer than the horizontal sides, causing the hexagons to be taller than they are wide.

Cone cells show a subdivision similar to that seen in secondary pigment cells. Polar and equatorial cone cells are of the "horizontal" type, bridging equivalent receptors R3 and R4, and R1 and R6 respectively. Polar and equatorial cone cells also contact both primary pigment cells as well as the horizontal secondaries. Anterior and posterior cone cells contact non-equivalent photoreceptors, only a single primary pigment cell, and two adjoining oblique secondaries. During development, the anterior and posterior cone cell nuclei rise apically in paired fashion and are then followed by the paired outward movement of the polar and equatorial cone cell nuclei.

The two primary pigment cells are clearly equivalent cells reflected across the dorso-ventral axis of the ommatidium and contact equivalent partners including both horizontal and oblique secondary pigment cells and cone cells.

Consideration of the contacts made by photoreceptors with cone cells and secondary pigment cells in terms of "horizontal" and "oblique" subtypes indicates that R1, R3, R4 and R6 interact equally with both types, while R2 and R5 contact only oblique cells.

An alternate perspective on the "horizontal/oblique" distinction can be seen in the connectedness of retinal cells between ommatidia. Each of the shared secondary pigment cells associates specifically with two outer photoreceptors in two neighboring ommatidia. The horizontal pigment cell "family" includes R1,R6/R3,R4, while oblique cells include either R2,R3/R5,R6 or R1,R2/R4,R5. From the symmetry-derived equivalences R1=R6, R2=R5, and R3=R4, it is apparent that the two oblique families are interchangeable and not equivalent to the horizontal family. It is notable that the seamless meshing of the chiral ommatidial forms across the equator is mediated by cell-cell associations that preserve normal family relations. The specificity of the secondary pigment cell-photoreceptor contacts suggests a plausible means by which autonomously formed ommatidia could be brought into register.

Several salient features of ommatidial construction are encountered in precise molecular assemblies such as enzyme complexes and viral capsules. The considerations of accuracy, economy, and stability that favor symmetrical self-assembly in these molecular aggregates may make an analogous process the method of choice for building an ommatidium.

CELL IDENTITY AND OMMATIDIAL ASSEMBLY

Some form of cell recognition that allows cells to discriminate among neighbors seems to be required to account for the in-place, close-grained specification of distinct cell types in the retinal epithelium and for the precision of cell contacts in the adult eye. As a speculative venture, it can be proposed that a simple combinatorial code endows retinal cells with the identities used to build an ommatidium (Fig. 9). The code is an abstraction of the contacts and symmetries of the eye. Cells are postulated to display a cell-surface marker that connotes cell type, plus one or both of two other markers that indicate two additional qualities, "horizontal" and "oblique". It is important to note that although

Fig. 9. Hypothetical code for ommatidial cell identity. Cell types, concentrically from inside out: photoreceptors R1-R6, cone cells, primary pigment cells, secondary pigment cells. "Horizontal" cells: white. "Oblique" cells: black. "Horizontal/oblique" cells: checked.

these names are derived from the eye axes, the qualities they connote are not anatomical; a cell displaying the quality "horizontal" need not be oriented in a specific direction, nor is there a contradiction in a cell simultaneously displaying horizontal and oblique markers. In this scheme, the code for either an anterior or a posterior cone cell would be "cone cell/oblique". The code for R1,R2,R3, and R4 is "photoreceptor/horizontal/oblique", while R2 and R5 are "photoreceptor/oblique". The unique nature of the unpaired central cells makes their inclusion in this scheme problematic, but R7 should be "photoreceptor/horizontal". Ommatidial cells are postulated to associate preferentially with neighbors with which they share markers in common. For example ,the characteristic grouping of R1,R6,R3, and R4 with the horizontal secondary pigment cell would be mediated by their common "horizontal" markers. The eye is an interlocking network of cells stitched together by overlapping, shared properties

In addition to supplying the markers needed to refine the contacts made by retinal cells, the proposed code could provide the identities read by cells during development in order to determine their appropriate cell type. If the walls of the niches formed by cells already incorporated into the precursor are painted with these markers, an undetermined cell filling the niche could use a simple logic to interpret its position and determine its proper role. A simple cell identity code may suffice in the developing retina since cells are determined in place and do not need to sort out to form ordered assemblies.

While hypothetical, the model serves as a framework for thinking about ommatidial assembly and makes testable predictions. It predicts that markers for cell identities may be few in number and distributed in unanticipated ways. For example, a monoclonal antibody (mAbs) to a "horizontal" marker would be found on all cell types, but excluded from anterior and posterior cone cells as well as R2 and R5 and the oblique secondaries. We are currently examining cell surfaces of developing Drosophila eyes using mAbs.

MONOCLONAL ANTIBODIES AND EYE DEVELOPMENT

Monoclonal antibodies (mAbs) are well suited to examining cell surfaces, and we have been using them to look at the changing complexions of retinal cells during development. Monoclonal antibodies to antigens of the developing eye are readily generated using conventional hybridoma techniques. Mice are immunized with crude membrane preparations from either developing or adult eyes, and mAbs that detect antigens expressed during eye development are easily selected using immunostaining of whole mounts of the retinal epithelium (Fig. 10). The reiterated, temporal series of developing ommatidia displayed in the retinal epithelium is particularly useful for selecting antibodies of interest; antigens appearing on ommatidial cells paint a "wallpaper" pattern starting at distances from the morphogenetic furrow that correspond to the time of maturation when the antigen is expressed.

Two antigenic changes on the surfaces of developing retinal cells revealed by mAbs are shown in fig. 10.(a) & (b). The disc in fig. 10a was stained with mAb 302, an antibody that stains retinal

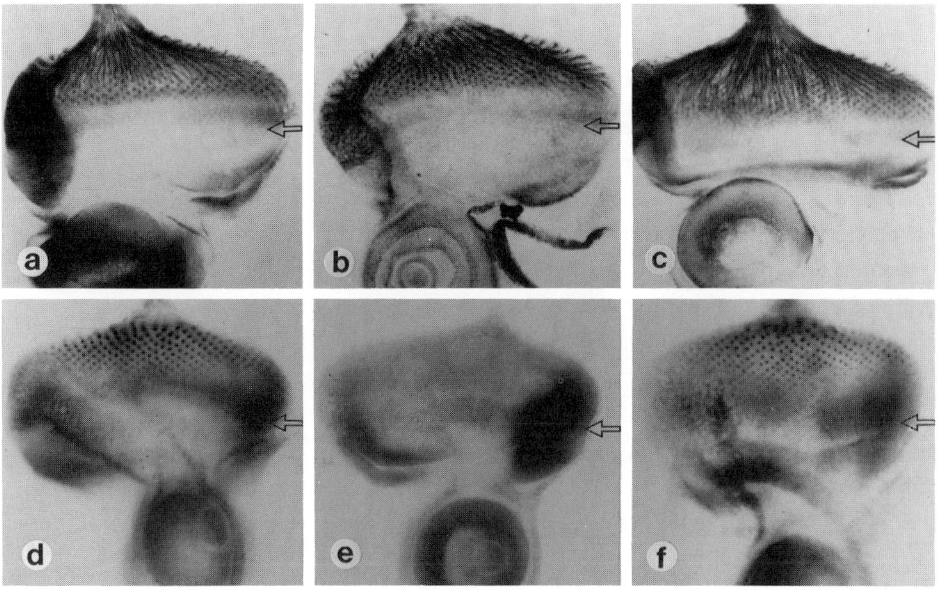

Fig. 10. Eye discs (retinal epithelia) stained with two different monoclonal antibodies. In each panel, the posterior end of the disc is marked by the optic stalk and the location of the morphogenetic furrow is marked with an arrow. The antennal disc, an epithelial primordium for the antenna is attached to the anterior of the eye disc, and is the concentric-ringed structure at the bottom of each panel. Top row shows staining of (a) normal (b) glass^{3} and (c) lozenge^{50e} with mAb 302, an antibody that stains photoreceptor cell bodies and axons. Each dot corresponds to an ommatidial precursor. Bottom row shows staining of (d) normal (e) glass^{3} and (f) lozenge^{50e} with mAb 301, revealing the apical tips of the photoreceptors, each dot again corresponds to an ommatidial precursor. Note mAb 301 staining is absent from glass eye disc. (From Lebovitz and Ready, in preparation)

cells starting at the furrow and prominently highlights photoreceptors and their axons beginning about the symmetrical cluster stage. The disc in fig.10d was stained with mAb 301, an antibody that is specific for photoreceptors and displays intense staining at the apical tips of the cells. The 301 antigen appears on receptors approximately 12-18 hours after the initial formation of the five-cell precursor and shows a strong increase in staining with maturation. In adult eyes, the antigen is highly concentrated in all eight rhabdomeres and is not found on cells other than photoreceptors. Our laboratory, as well as Benzer's (Zipursky et al., 1984) have isolated additional mAbs that reveal still other patterns of changing antigen expression in the developing eye. None of the antibodies isolated to date subdivide ommatidia in the manner suggested by the hypothetical identity code.

Identifying differentiation antigens using monoclonal antibodies is relatively straightforward; interpreting their significance is another matter. It is likely that many of the cell-surface changes that accompany differentiation will be related to the particulars of the cell's specialized function rather than the mechanisms of pattern formation. It would be useful to manipulate antigens in order to test ideas about possible roles for antigens and mutations offer valuable tools for this purpose. In principle, it should be possible to genetically modify or delete a specified antigen via mutation of its structural gene, and observe the consequences of such a lesion on development. In practice, this has yet to be achieved, but we have made a foray into this area by screening a sample of eye mutants against our monoclonals to see whether existing eye mutants affect the expression of eye differentiation antigens.

We chose six mutants with defects ranging from severe to minimal eye disruption and tested them for staining with mAbs 301 and 302. Antibody stained eye discs of normal flies and two of the mutants, glass^{3} and lozenge^{50e} are shown in fig. 10. All of the mutants except glass^{3} showed normal expression of both antigens. In glass^{3} eye discs, mAb 302 reveals photoreceptors organized into normal-looking clusters in a somewhat disordered array (fig. 10b). However, mAb 301 staining is strikingly absent (fig. 10b). Two days later, when normal photoreceptors begin to elaborate rhabdomeres, glass^{3} photoreceptors degenerate, leaving the mutant eye a shell of cuticular lenses over pigment cells and debris. The restricted distribution of the 301 antigen is paralleled by the tissue-specific action of glass; glass mutants do not show defects in tissues other than the retina.

The defect that causes the absence of the 301 antigen and subsequent photoreceptor degeneration is not known, but it is noteworthy that the developmental crisis that kills mutant photoreceptors occurs when abundant production of the 301 antigen would be required. Glass mutants also lack another photoreceptor-specific antigen, 24B10, independently isolated by Benzer and his co-workers (Fujita et al., 1982). The pattern of staining with mAb 24B10 has many features in common with that of 301, but subtle staining differences suggest the antibodies detect distinct aspects of the cell surface. The structural gene for the 24B10 antigen has been located far from the position of glass (Ziprusky et al., 1985). Glass may thus participate in a program of photoreceptor differentiation; arrest of the program in the mutant could block the expression of several photoreceptor antigens.

As with any lesion-based approach, mutant analysis is subject to the caveat that eye defects result from developmental machinery operating with a missing or damaged element; what is lacking is not necessarily the normal role of the defective gene. Coupling monoclonal antibodies to a mutant analysis adds the additional complication that the absence of mAb staining in a mutant means simply that a particular determinant is missing; the remainder of the macromolecule, including its functional domains may be intact and functioning normally.

Despite these reservations, the sensitive structural probes available through monoclonal antibody and recombinant DNA techniques offer a new, ultra-high resolution view of eye development, and many fundamental problems of eye development are within the reach of these methodologies. What signals in the local environment do cells read to determine their proper roles, and how are they read? What genes participate in ommatidial assembly and how is their action coordinated? What molecules mediate the physical interactions between cells that arrange them into precise assemblies? The goal, and major hurdle, of this approach is to find the congruence between the molecular and cellular pictures of eye develoment.

EPILOGUE

The algorithm of ommatidial assembly is written in both genetic and epigenetic terms. While pieces of the process may be isolated for close scrutiny, they are not used piecemeal in development; each is an integral element of a finely balanced program. An understanding of the how the ommatidial pattern arises will of necessity include a description of how the distinct geometries of the retinal epithelium selectively activate genes as well as how selective gene expression is converted into distinct geometries. The special feature of the Drosophila eye is that both aspects of development are readily accessible.

ACKNOWLEDGEMENT

This work was supported by funds from the Sloan Foundation and the National Science Foundation.

REFERENCES

Becker, H.J. (1957). Uber Rontgenmosaikflecken und Defektmutationen am Auge von Drosophila und die Entwicklungphysiologie des Auges. Z. Induk. Abst. Vererb. Lehre. 88, 333-373.
Bodenstein, D. (1953). Postembryonic development. In Insect Physiology (ed. K.D. Roeder), pp. 822-865. New York: Wiley.
Braitenberg, V. (1967) Patterns of projection in the visual system of the fly. I. Retina-lamina projections. Exp. Brain Res. 3, 271-298.
Dietrich, W. (1909). Die Facettenaugen der Dipteren. Zeitschr. f. wiss. Zool. 92, 465-539.
Foster, G.G., and Suzuki, D.T. (1970). Temperature-sensitive mutations in Drosophila melanogaster, IV: A mutation affecting eye facet arrangement in a polarized manner. Proc. Nat. Acad. Sci. 67, 738-745.
Fristrom, D.K., and Fristrom, J.W. (1982). Cell surface binding sites for peanut agglutinin in the differentiating eye disc of Drosophila. Dev. Biol. 92, 418-427.
Fujita, S.C., Zipursky, S.L. Benzer, S., Ferrus, A. and Shotwell, S.L. (1982). Monoclonal antibodies against the Drosophila nervous system. Proc. Natl. Acad. Sci. 79, 7929-7933.
Harris, W.A., Stark, W.S. and Walker, J.A. (1976). Genetic dissection of the photoreceptor system in the comound eye of Drosophila melanogaster. J. Physiol. (Lond.) 256, 415-439.
Hyde, C.A.T. (1972). Regeneration, post mbryonic induction and cellular interaction in the eye of Periplaneta americana. J. Embryol. Exp. Morph. 27, 367-369.
Lawrence, P.A., and Green, S.M. (1979). Cell lineage in the developing retina of Drosophila. Dev. Biol. 71, 142-152.
Lawrence, P.A., and Shelton, P.M.J. (1975). The determination of polarity in the developing insect retina. J. Embryol. exp. Morph. 33, 471-486.
Lebovitz, R.M. and Ready, D.F. (1982). Monoclonal antibodies to developing Drosophila retina. Neurosci. Abst. 8, 702.
Meyerowitz, E.M., and Kankel, D.K. (1978). A genetic analysis of visual system development in Drosophila melanogaster. Dev. Biol. 62, 112-142.
Nowel, M.S., and Shelton, P.M.J. (1980). The eye margin and compound-eye development in the cockroach: evidence against recruitment. J. Embryol. Exp. Morphol. 60, 329-343.
Poodry, C.A., Hall, L., and Suzuki, D.T. (1973). Developmental properties of <u>shibire</u>^{ts1}, a pleiotropic mutation affecting adult and larval locomotion and development. Dev. Biol. 32, 373-386.
Ready, D.F., Hanson, T.E., and Benzer, S. (1976). Development of the Drosophila retina, a neurocrystalline lattice. Dev. Biol. 53, 217-240.
Ready, D.F., Goldsmith, J., and Ecker, J. (1985). How the fly's eye finds the fly's brain. Soc. Neurosci. Abstracts.
Shelton, P.M.J., Pfannenstiel, H.-D., and Wachmann, E. (1983). Regeneration of the eye margin in Periplaneta americana (Insecta, Blattodea). J. Embryol. Exp. Morphol. 76, 9-25.

Trujillo-Cenoz, O., and Melamed, J.(1966) Compound eye of Dipterans: anatomical basis for integration-an electron microscope study. J. Ultrastruct.Res., 16 395-398.

Tomlinson, A. (1985). The cellular dynamics of pattern formation in the eye of Drosophila . J. Embryol. exp. Morph., in press.

Tomlinson, A. and Ready, D.F. (1985) sevenless, a cell-specific homoeotic mutation of the Drosophila eye. (submitted).

Waddington, C.H. (1962). New Patterns in Genetics and Development. Columbia University Press.

White, R.H. (1961). Analysis of the development of the compound eye in the mosquito, Aedes Aegypti. J. Exp. Zool. 148, 223-240.

Wolsky, A. (1949). The analysis of eye development in insects. Trans. N.Y. Acad. Sci. Ser. 11, 18, 592-596.

Zipursky, S.L., Venkatesh, T.R.,Teplow, D.B. and Benzer, S. (1984). Neuronal development in the Drosophila retina: monoclonal antibodies as molecular probes. Cell 36, 15-26.

Zipursky, S.L. Venkatesh, T.R. and Benzer, S. (1985). From monoclonal antibody to gene for a neuron-specific glycoprotein in Drosophila. Proc. Natl. Acad. Sci. 82, 1855-1859.

A Minimum Diameter Limit for Retinal Rod Outer Segment Disks

Joseph M. Corless

The photoreceptive rod outer segments (ROSs) of the frog retina are roughly cylindrical in shape, and measure approximately 55 μm in length and 5-8 μm in diameter (figure 1 A). Contained within the plasma membrane of the ROS is a well-ordered, axial array of flattened, unilamellar membranous vesicles called disks. The axial flattening of each disk is associated with two characteristic membrane profiles in longitudinal thin sections through the ROS (figure 1 B). One observes 1) an extensive, central lamellar domain (L), in which the apposing membranes lie parallel and slightly separated, and 2) a highly curved domain at the edge of the disk, called the terminal loop (TL), which interconnects the two lamellar membrane regions.

In cross-sectional views, the disks are not simply circular in outline (figure 1 C). Each disk displays a series of indentations along its perimeter, termed incisures (I), which give the disk a partially sectored or lobulated appearance. Thus, the perimeter or edge of each ROS disk has two distinctive components: 1) incisural elements (I), approximately radial in organization, and 2) marginal segments (M), which describe a roughly circular profile.

GENERAL ORGANIZATION OF THE TERMINAL LOOP COMPLEX

Three special features of the disk perimeter have focused our attention on the TL region. 1) The margins and incisures of adjacent disks are axially aligned. What are the structural and molecular bases for the connectivity implied by such alignments? 2) The shapes of ROS disk perimeters differ considerably among vertebrate species (figure 2). What is the significance of these

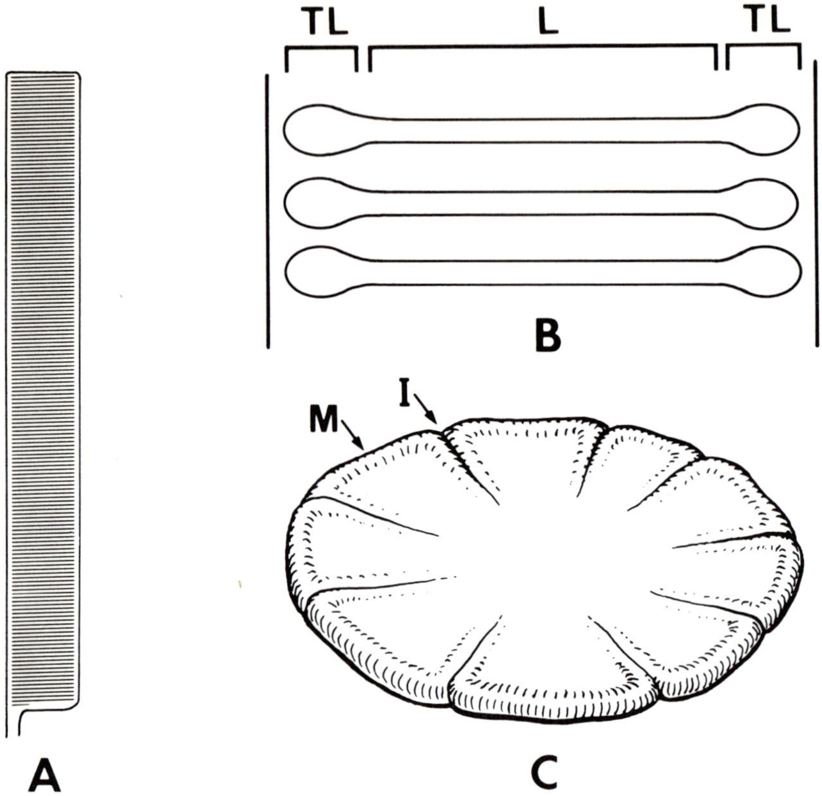

Figure 1. Morphological features of frog retinal rod outer segments.
A. Longitudinal section through the ROS, indicating the closely spaced array of photoreceptive disks enclosed within the ROS plasmalemma.
B. Enlargement of three disks, showing the lamellar (L) and highly curved terminal loop (TL) regions of the disk membranes (not to scale) . The axial spacing of frog ROS disks is approximately 295 Å (Corless, 1972).
C. Three-dimensional representation of a single disk. The perimeter of the disk is folded, and consists of marginal (M) and incisural (I) components. Drawing by Mr. Brad Smith, after Rohlich (1981).

variations, and what might they imply about the principles of disk perimeter construction? 3) Disk margin formation appears coupled with a process which progressively isolates disk precursor components from the ROS plasmalemma (Steinberg et al., 1980). What are the molecular mechanisms and structural correlates of this process? Our initial approach to these questions has been to acquire a more detailed structural understanding of the TL region.

The perimeter of frog ROS disks is characterized by a clustering of specific morphological elements called the terminal loop complex (Corless, 1983; Corless et al., 1985). This complex appears to be a repeating structural unit organized along the entire length of the disk perimeter. Figure 3 illustrates the disk incisure pattern from one cross-sectional image. In region A of figure 3, two incisures are shown in greater detail. Continuity between marginal and incisural components is more clearly indicated. Recessed about 200 Å from the edge of the disk, one observes a sequence of discrete densities with a periodic spacing of about 175 Å.

The inset provides a schematic longitudinal view of the area boxed in figure 3 A. The TL complex consists of (a) semi-circular to crescent-shaped densities within the TL loop, (b) linear interdisk densities spanning the cytoplasm between adjacent disks, and (c) distinctive freeze-fracture particles located between (a) and (b). The TL complex appears to have an overall dimeric organization in that the upper and lower halves of the complex seem related by a 2-fold axis (arrow). These structural elements of the TL complex appear capable of maintaining the high curvature of this region, as well as the alignment of disk margins and incisures. The overall morphology of the TL complex probably reflects the spatial distribution of the large TL protein localized along the ROS disk perimeter by Papermaster et al. (1978) (Corless et al., 1985).

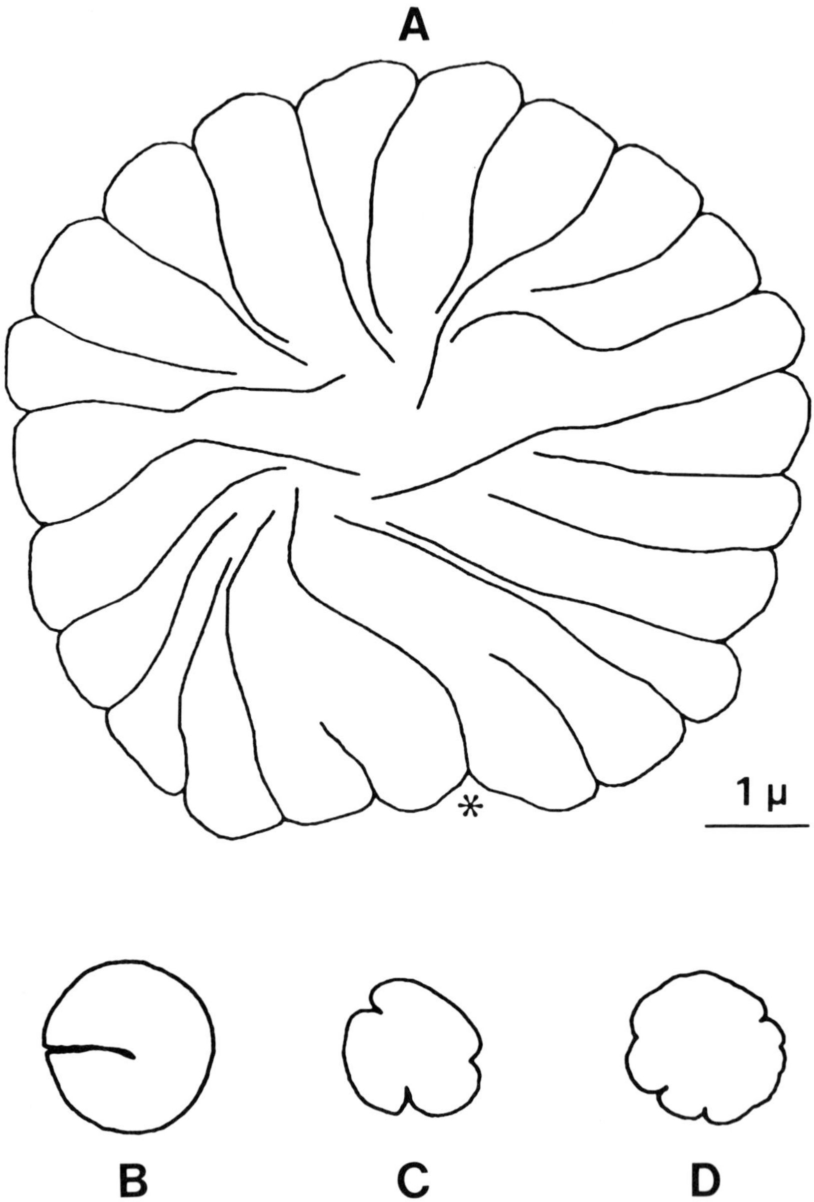

Figure 2. Perimeter patterns observed in selected ROS disks. (A) Frog, (B) mouse, (C) cat, and (D) human. An asterisk (*) indicates the location of the ROS extension of ciliary cytoplasm. Scale bar: 1μ. See table II for literature references.

PROPOSED ROLE OF THE TERMINAL LOOP COMPLEX IN ROS DISK DEVELOPMENT

We have developed a general hypothesis in which components of the TL complex serve as the morphogens which locally direct the growth and differentiation of the disk margin (Corless, 1983; 1984; Corless & Fetter, 1985). In this view, the resultant structure of the TL region reflects in some fashion the process by which it was constructed. One novel prediction derived from the model is that there is a minimum size limit for ROS disks. In the following sections, we will summarize these arguments, derive quantitative predictions, and compare such results with experimental data.

A MINIMUM DIAMETER LIMIT FOR ROS DISKS

The prediction of a minimum diameter requirement is derived from the simultaneous constraints imposed by the architecture of the disk perimeter, the lamelliform geometry of the disks, and their biochemical composition. The overall mechanism of ROS disk formation solves two problems. Firstly, it produces extended lamellar domains composed of lipid and rhodopsin, whose primary function is the absorption of light. Secondly, the disks become detached from the ROS plasmalemma. The resulting perimeter of the disk is characterized by the periodic arrangement of TL complexes. We will go one step further and postulate that the disk perimeter must be built in this fashion. Thus, the rod photoreceptor cell must synthesize 1) TL components to construct the perimeter of the disk, and 2) rhodopsin and lipids to construct the central lamellar domains. These components are manufactured by the cell and then incorporated into the ROS in specific molecular ratios. Accordingly, we can conceptualize a specific lamellar area per TL complex, in both a geometrical and a biochemical sense.

The value of this ratio can be estimated in the case of frog red ROSs (consult figure 3). We have measured the cross-sectional lamellar area and perimeter length of red ROS disks (Corless et al., 1985). If we divide the perimeter length by the repeat period of the TL complex (175 Å), we obtain an estimate of the number of TL complexes contributing to the perimeter. Dividing the lamellar area by this number, we obtain A_o = 514,458 $Å^2$, i.e., the cross sectional lamellar area per TL complex. This calculation assumes that all TL complexes are localized along the disk perimeter, and that none are distributed within the lamellar domains. As an initial approximation, this is quite reasonable for frog red ROSs.[1] We have argued that the large TL protein described by Papermaster et al. (1978) is an integral component of the TL complex (Corless et al., 1985). Papermaster et al. (1978) have shown that this TL protein is confined almost exclusively to the disk perimeter in red ROSs.

The cross-sectional lamellar area per TL complex may be given a geometrical representation. Consider a flattened single, unilamellar vesicle composed of the number of rhodopsins and lipids present per TL complex. Such a construction is illustrated in figure 4 (a), and may be viewed as the "morphological unit" associated with the "compositional unit."

The disk shown in figure 4 (a) has an area equal to A_o. It will also have a minimum perimeter (P_L) determined by the lamellar area: $P_L = \sqrt{4\pi A_o}$. From table I, we note that P_L = 2543 Å. However, the maximum length of normal disk perimeter (P_{TL}) constructable in this case is only 175 Å, the length determined by one TL complex. Therefore, our compositional unit cannot be structured as a normal ROS disk: there are insufficient TL complexes to match the minimum perimeter defined by the lamellar domains. How does the rod cell overcome this apparent constraint? Very simply. It merely increases the number of compositional units incorporated per disk. If n such units are included, the maximum length of normal perimeter

Figure 3. Organization of the terminal loop (TL) complex in frog ROSs. The perimeter pattern of a frog disk is shown. In region A, the appearance of the TL complex in planar views is indicated, and consists primarily of a row of discrete densities spaced approximate 175 Å apart, and recessed approximately 200 Å from the disk edge. This substructure is observed along both marginal and incisural components of the disk perimeter. In longitudinal views through the TL region (upper inset, cf. figure 1B), other densities and components are present. See text for further details. The density labeled (b) in the inset corresponds to the discrete densities observed in planar section.

---

[1]For frog ROS disks, we estimate the number of TL complexes in the perimeter to be 8438. For toad ROS disks, which are of comparable size, Roof & Heuser (1982) have described a system of axially oriented rim filaments. These appear to correspond to our linear interdisk densities (figure 3(b)), and number approximate 8000. Thus, our estimates for the number of repeating units along the disk perimeter are similar.

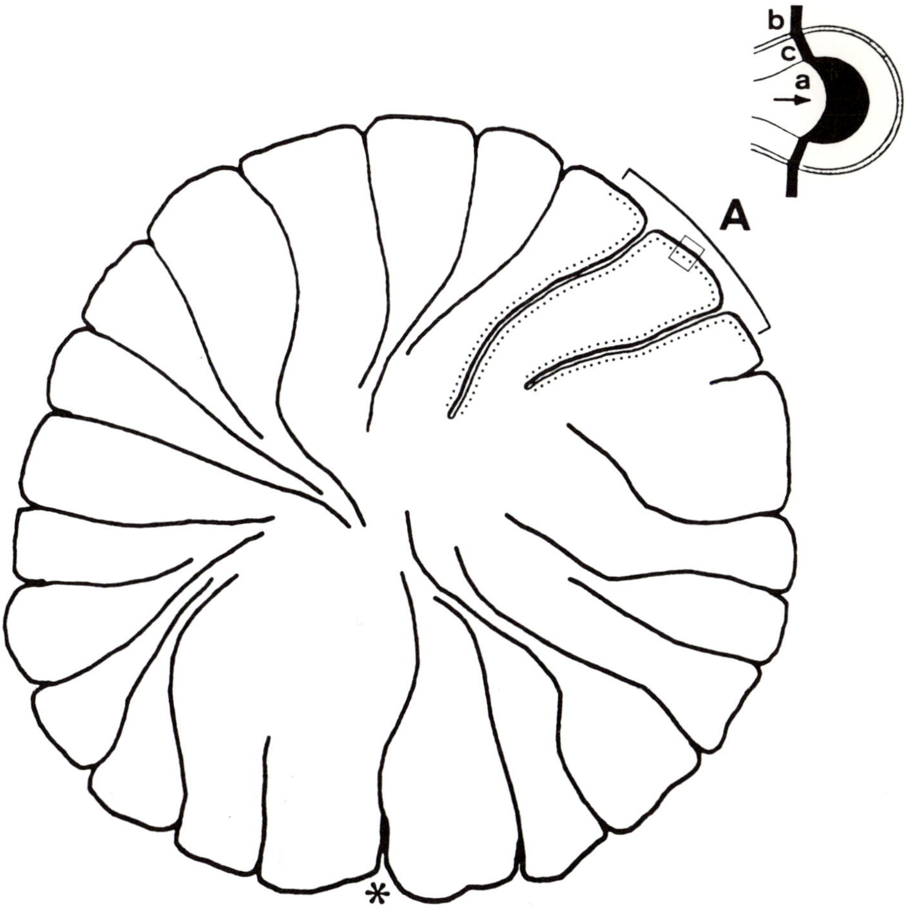
b
c
a
A
*

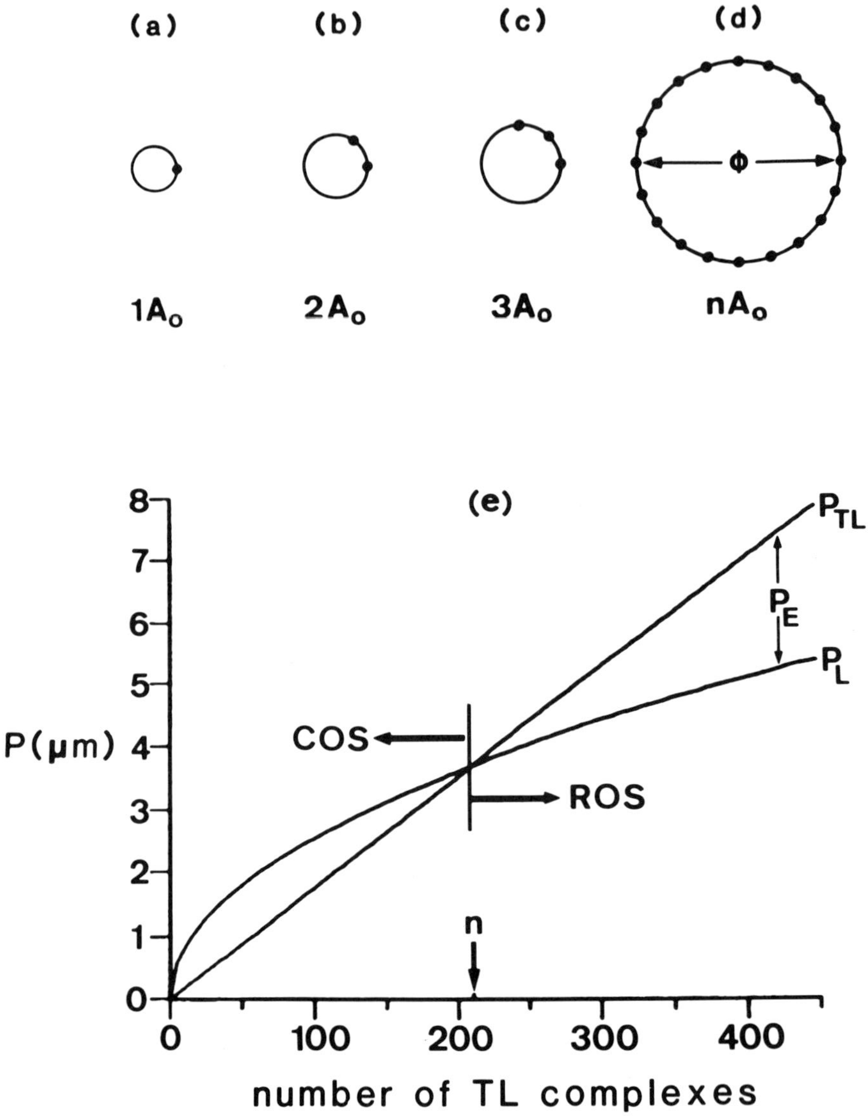

Figure 4. Geometrical and graphical representations of minimum disk diameter prediction. See text and tables I and II for explanation. In (a-d), each dot on the perimeter represents a TL complex.

TABLE I

Evolution of Morphological Parameters for Minimum ROS Disk Prediction

Parameter	Calculated Values			
n	1	2	3	211
P_L (Å)	2543	3596	4404	3.69×10^4
P_{TL} (A)	175	350	525	3.69×10^4
P_L/P_{TL}	14.53	10.27	8.39	1.00
∅ (μm)	0.081	0.114	0.140	1.18

DEFINITIONS

n = number of TL complexes = number of compositional units
d = spacing of TL complexes along disk perimeter = 175 Å
A_o = lamellar cross-sectional area per TL complex = 514,458 $Å^2$
P_L = minimum perimeter determined by lamellar area = $\sqrt{n} \times \sqrt{4\pi A_o}$
P_{TL} = maximum perimeter length constructable from n TL complexes = n x d
∅ = diameter of lamellar domain; approximately equal to ROS

constructable is $P_{TL} = n \times d$, where d = 175 Å. Whereas P_{TL} increases linearly with n, the minimum perimeter determined by the lamellar domains increases only by $\sqrt{n}$, i.e., $P_L = \sqrt{n} \times \sqrt{4\pi A_o}$. A normally constructed ROS disk perimeter will be possible only when $P_{TL} \geq P_L$. Figure 4 (e) charts the values of P_{TL} and P_L as a function of the number of compositional units per disk. When n $\geq$ 211, $P_{TL} \geq P_L$. Thus, only after n reaches a value of 211 does it become possible to construct a normal disk perimeter along the circumference defined by the lamellar domains. This process is illustrated in figure 4(a - d). From table I, this minimum disk diameter is 1.18 μm. At smaller values of n, our model of disk formation would predict that closure of the disk margin cannot occur. The developing disk would still be continous with the ROS plasmalemma, and would have a topology characteristic of frog cone outer segment (COS) disks. Thus, the disk must achieve a minimum diameter of 1.18 μm to acquire the minimum number of TL complexes necessary for construction of a closed perimeter. In a more general description, we might say that the rod cell uses a scaling approach in the construction of its outer segment disks (Schmidt-Nielsen, (1984).

The minimum diameter predicted from the structural features of frog ROS disks is extremely close to the smallest ROS (and disk) diameters observed in vertebrate retinas. For example, human ROSs measure approximately 1.0-1.5 μm in diameter (Cohen, 1965; Hogan et al., 1971), rhesus monkey 1.2-1.5 μm (Young, 1971; Cohen, 1961), macaque approximately 1.6 μm (Cohen, 1961; 1965), rat approximately 1.7 μm (Hagins et al., 1970), cat 1.3-1.4 μm (Steinberg & Wood, 1975), and guinea pig 2.2-2.3 μm (Sjöstrand, 1949). Mouse ROSs are 1.4-1.5 μm in thin sections (Cohen, 1960; Carter-Dawson & LaVail, 1979), but slightly larger (approximately 1.6 μm) by freeze-fracture (Andrews & Cohen, 1979).

Implicit in these comparisons are the assumptions 1) that the lamellar membrane area per TL complex is the same for all species, and 2) that the spacing of TL complexes along disk perimeters is constant. Available data suggest that these assumptions are at least reasonable first approximations. Regarding the first point, we note that the total lipid/total protein ratio (wt/wt) of vertebrate ROS disk membranes in general is about 1 (Fliesler & Anderson, 1983). If we compare, for example, the recent data for frog and rat disk membranes, the corresponding protein/lipid and phospholipid/total lipid ratios are almost identical. Converse (1979) has emphasized the similarities in molecular weight and relative abundance for the large intrinsic membrane proteins in frog, cattle, rabbit and human outer segments. Thus, a common value for A_o, the lamellar cross-sectional area per TL complex, seems a reasonable approximation. Similarly , a constant TL complex spacing also seems reasonable, considering that all ROSs encounter equivalent problems of disk isolation and perimeter construction. Therefore, it seems unlikely that the close correspondence between the predicted minimum disk diameter and experimentally observed values is merely coincidental.

It should be emphasized that the conceptual requirement of a minimum disk diameter limit applies to each rod cell individually. The value of A_o will be determined by the ratios of membrane components synthesized and incorporated into the disk membranes of each rod cell. Our theory does <u>not</u> predict a maximum size limit for either rod or cone outer segment disks.

THE SIGNIFICANCE OF EXCESS DISK PERIMETER

Rod cells do not construct circular disks of minimum diameter. This point is evident in figure 2, where the disk perimeter is always in excess of the minimum length needed simply to interconnect the lamellar domains. Formally, we interpret this geometrical feature with reference to figure 4 (e). If all the TL complexes can be utilized to build additional disk perimeter, then for values of n>211, the constructable perimeter (P_{TL}) will always exceed by P_E the minimum perimeter defined by the lamellar area (P_L). In order for the perimeter of the lamellar domain to remain congruent with the perimeter length defined by the TS complexes, the shape of the perimete must fold or buckle: it cannot remain circular. Thus, the incisures of ROS disks are viewed as a consequence of the continuing perimeter construction which accompanies disk maturation. The mechanism proposed for this process is discussed elsewhere (Corless & Fetter, 1985).

If this view is substantially correct, then there are two ways to estimate the number of TL complexes for any ROS disk. The number based upon the lamellar area of a disk would be:

$$N_L = A_L / A_o$$

where A_L is the area of the lamellar domain, and A_o is the lamellar area per TL complex. The estimate based upon perimeter length (P) is given by

$$N_P = P / d ,$$

where d = 175 Å, the spacing of TL complexes.

Table II indicates the values of N_L and N_P for several species of ROS disks found in the literature. In general, the values of N_L and N_P show reasonably good agreement, especially for the mouse ROS prepared by freeze-fracture techniques. For aldehyde / OsO_4 - fixed specimens, the correspondence between N_L and N_P is not quite as close. When only OsO_4 is used, the values of N_L and N_P diverge even further. Although the sample size is small, the results tend to suggest that the method of fixation affects estimates of N_L to a greater extent than estimates of N_P. Such a trend could be expected, because the degree to which organic solvents extract lamellar components will depend upon the effectiveness with which such components have been cross-linked (Gigg & Payne, 1969; Nir & Hall, 1974; Kalina & Pease, 1977). The structure of the disk perimeter appears more resistant to extraction procedures (Falk & Fatt, 1969; Doggenweiler, 1972).

The prediction of excess disk perimeter can be developed in a more graphical way. If we only measure the lamellar area of a disk, we can estimate the maximum disk perimeter: $P_{TL} = N_L \times d$. If we inspect the disk cross-section and count the number of incisures

TABLE II: Morphometric Data for Selected Vertebrate ROS Disks

Species	Figure 2	Lamellar Area (A_L μm^2)	Perimeter (P μm)	N_L	N_P	N_L/N_P	Reference & Fixation Procedure
frog	a	43.41	147.66	-	8438	-	b) aldehyde/ tannic acid/ OsO_4
mouse	b	1.99	6.82	387	390	0.99	c) aldehyde/ glycerol/ freeze-fracture
cat	c	1.31	4.88	255	279	0.91	d) aldehyde/ OsO_4
human		1.37	5.39	267	308	0.87	e) aldehyde/
	d	1.66	5.29	323	302	1.07	OsO_4
		1.22	4.97	218	284	0.77	f) OsO_4
		1.06	5.00	206	86	0.72	g) Os

a) Disk perimeter and area were measured from micrographs using a Numonics Model 1224 Electronic Digitizer (Numonics Corp., Lansdale, PA)
b) Corless et al., 1985. Average values. Preparation: 30 min 1% glutaraldehyde with 0.2% tannic acid, 5-20 min 1% OsO_4, ethanol, propylene oxide, Epon 812.
c) Andrews & Cohen, 1979 (their figure 5). Preparation: $\geq$ 12 hr 2.5% glutaraldehyde, 1-2 hr buffered 25% glycerol, frozen in Freon 2 near -160°C.
d) Steinberg & Wood, 1975 (their figure 1B). Preparation: 2-3 hr 3% paraformaldehyde/glutaraldehyde, 2-3% hr 2% OsO_4, $\geq$ 2 hr uranyl acetate,
e) Hogan et al., 1971 (their figure 9-14; see Spitznas & Hogan, 1970). Preparation: 2 hr 2 % glutaraldehyde and 1 % paraformaldehyde, 2 hr OsO_4, 2 hr uranyl acetate, acetone, propylene oxide, araldite.
f) Cohen, 1965 (his figure 6).
g) Cohen, 1969 (his figure 18 C). Preparation: 1 hr 2 % OsO_4, ethanol, toluene, araldite.

present, we can add a general shape constraint to our perimeter prediction. Using these three parameters, (lamellar area, number of incisures, and predicted perimeter length), one can devise simple algorithms to generate shapes which incorporate these values. Figure 5 (b-d) illustrates one series of symmetrical shape predictions for the disks shown in figure 2 (b-d). For mouse and cat, the predicted perimeter and superimposed shape appear to closely model the observed disk geometries. The perimeter of the human ROS disk is more irregular in shape, so the symmetrical shape constructed for the predicted perimeter is less representative of the native pattern. The shape of the frog disk generated by the perimeter algorithm is included for comparison only; it is not "predicted" in the same sense as the other cases, because the frog morphometric data was used as the basis for the other predictions (table II).

In summary, our model of disk development predicts that ROS disks should have a minimum diameter limit. From morphometric analyses of frog ROS disks, the predicted minimum diameter is 1.18 μm. This value is in excellent agreement with the smallest diameters observed in vertebrate ROSs. The correspondence between prediction and observation provides support for our basic model of disk perimeter formation (Corless, 1983; 1984; Corless & Fetter, 1985). In the present context, our theory has been extended to address the apparent relationship between perimeter length and disk area in a quantitative way (cf. figure 2).

ACKNOWLEDGEMENTS

I am indebted to Mr. R.D. Fetter for critically reviewing this manuscript, to Mr. T.G. Scheider for technical assistance, and to Ms. Carol B. Atkins for unfailing secretarial support. This work was supported by U.S.P.H.S. Grant Nos. EY 01659 and No. 04922 from the National Eye Institute.

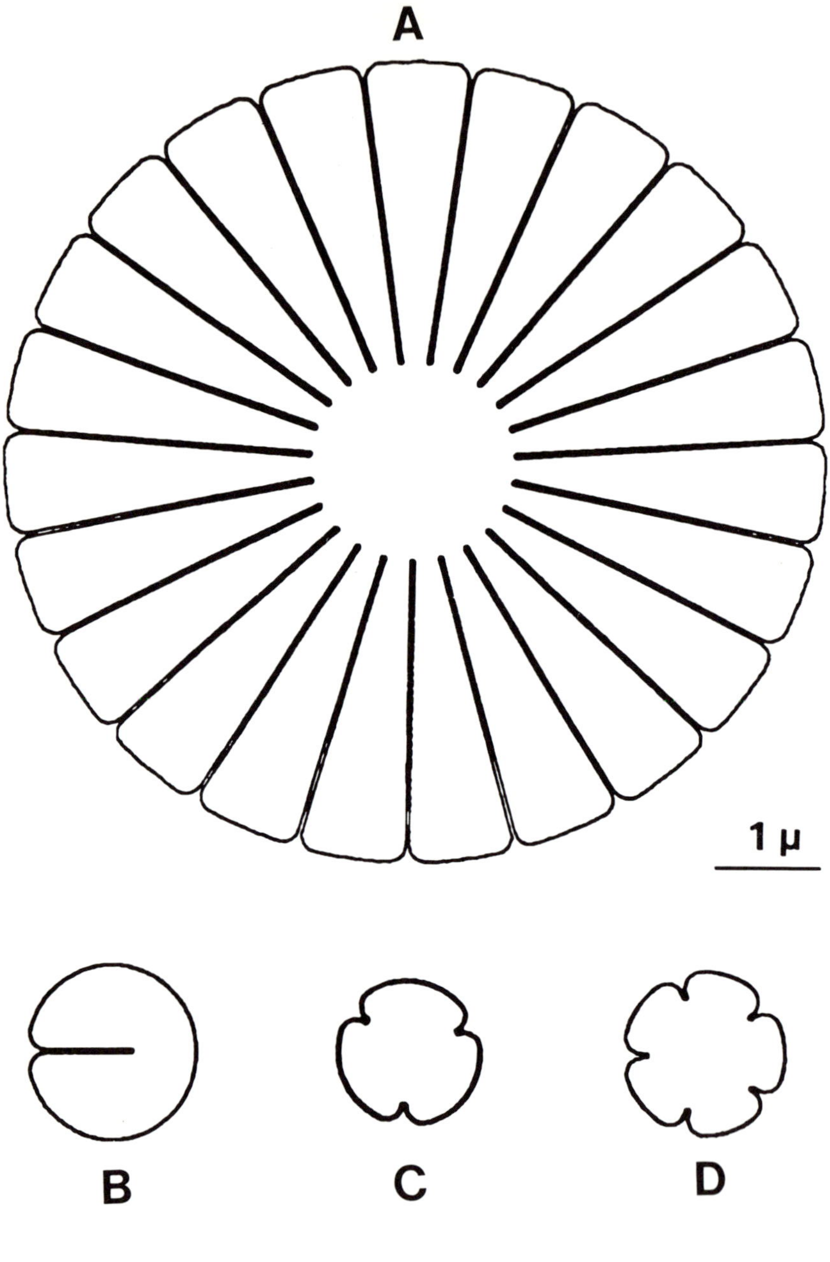

Figure 5. Prediction of perimeter length from disk area measurements. The calculated patterns correspond to the ROS disks illustrated in figure 2. (A) frog, (B) mouse, (C) cat, and (D) human. Scale bar: 1μ.

REFERENCES

Andrews, L.D. and A.I. Cohen. 1979. Freeze-fracture evidence for the presence of cholesterol in particle-free patches of basal disks and the plasma membrane of retinal rod outer segments of mice and frogs. J. Cell Biol. 81:215-228.

Carter-Dawson, L.D. and M.M. LaVail. 1979. Rods and cones of the mouse retina. I. Structural analysis using light and electron microscopy. J. Comp. Neurol. 188:245-262.

Cohen, A.I. 1960. The ultrastructure of the rods of the mouse retina. Am. J. Anat. 107:23-48.

Cohen A.I. 1961. The fine structure of the extrafoveal receptors of the rhesus monkey. Exp. Eye Res. 1:128-136.

Cohen, A.I. 1965. New details of the ultrastructure of the outer segments and ciliary connectives of the rods of human and macaque retinas. Anat. Rec. 152:63-80.

Cohen, A.I. 1969. Rods and cones and the problem of visual excitation. In The Retina: Morphology, Function and Clinical Characteristics. B.R. Straatsma, M.O. Hall, R.A. Allen and F. Crescitelli, editors. University of California Press, Berkeley. 31-62.

Converse, C.A. 1979. The large intrinsic membrane protein in rod outer segments: In vitro synthesis in cattle, and comparison in humans and rabbits. Exp. Eye Res. 29:409-416.

Corless, J.M. 1972. Lamellar structure of bleached and unbleached rod photoreceptor membranes. Nature (London) 237:229-231.

Corless, J.M. 1983. The terminal loop (TL) complex of frog retinal rod outer segment (ROS) disks: Implications for disk morphogenesis. Biophys. J. 41 (No. 2, Pt. 2): 124a.

Corless, J.M. 1984. The perimeter structure of photoreceptive lamellae in rod (ROS) and cone outer segments (COS): Implications for disk morphogenesis. Invest. Ophthalmol. Visual Sci. 25 (No. 3, Suppl.):65.

Corless, J.M. and R.D. Fetter. 1985. The terminal loop complex of frog retinal rod outer segment disk membranes. Implications for disk morphogenesis, membrane fusion and cell surface interactions. In preparation.

Corless, J.M., R.D. Fetter, O.B. Zampighi and M.J. Costello. 1985. Structural features of the terminal loop of frog retinal rod outer segment disk membranes. II. Organization of the terminal loop complex. In prepararation.

Doggenweiler, C.F. 1972. Effect of detergent treatment on photoreceptor membranes. Revista Latinoamericana de Microscopia Electronica 1:146-147.

Falk, G. and P. Fatt. 1969. Distinctive properties of the lamellar and disk-edge structures of the rod outer segment. J. Ultrastruct. Res. 28:41-60.

Fliesler, S.J. and R. E. Anderson. 1983. Chemistry and metabolism of lipids in the vertebrate retina. Prog. Lipid Res. 2:79-131.

Gigg, R. and S. Payne. 1969. The reaction of glutaraldehyde with tissue lipids. Chem. Phys. Lipids 3:292-295.

Hagins, W.A., R.D. Penn and S. Yoshikami. 1970. Dark current and photo-current in retinal rods. Biophys. J. 10:380-412.

Hogan, M.J., J.A. Alvarado and J.E. Weddell. 1971. Histology of the Human Eye. Philadelphia, W.B. Saunders Co. p.413

Kalina, M. and D.C. Pease. 1977. The preservation of ultrastructure in saturated phosphatidyl cholines by tannic acid in model systems and type II pneumocytes. J. Cell Biol. 74:726-741.

Nir, I. and M.O. Hall. 1974. The ultrastructure of lipid-depleted rod photoreceptor membranes. J. Cell Biol. 63:587-598.

Papermaster, D.S., B.G. Schneider, M.A. Zorn and J.R. Kraehenbuhl. 1978. Immunocytochemical localization of a large intrinsic membrane protein to the incisures and margins of frog rod outer segment disks. J. Cell Biol. 78:415-425.

Röhlich, P. 1981. Structure of retinal photoreceptor membranes as seen by freeze-fracturing. Acta Histochem. 23 (Suppl.): 123-136.

Roof, D.J. and J.E. Heuser. 1982. Surface of rod photoreceptor disk membranes: Intergral membrane components. J. Cell Biol. 95:487-500.

Schmidt-Nielsen, K. 1984. Scaling: Why is animal size so important? Cambridge, Cambridge University Press.

Sjöstrand, F. 1949. An electron microscope study of the retinal rods of the guinea pig eye. J. Cell. Comp. Physiol. 33:383-404.

Spitznas, M. and M.J. Hogan. 1970. Outer segments of photoreceptors and the retinal pigment epithelium. Arch. Ophthal. 84:810-819.

Steinberg, R.H. and I. Wood. 1975. Clefts and microtubules of photoreceptor outer segments in the retina of the domestic cat. J. Ultrastruct, Res. 51:397-403.

Steinberg, R.H., S.K. Fisher and D.H. Anderson. 1980. Disc morphogenesis in vertebrate photoreceptors. J. Comp. Neurol. 190:501-518.

Young, R.W. 1971. Shedding of discs from rod outer segments in the rhesus monkey. J. Ultrastruct. Res. 34:190-203.

The Role of Hormone-Stimulated cAMP Metabolism in Visual Cortical Plasticity

Chiye Aoki and Philip Siekevitz

Our interest has been to understand the biochemical basis of the developmental plasticity of vision. We have chosen to study the visual cortex of cats, because it is one of the systems that is most thoroughly characterized electrophysiologically and behaviorally. The critical period (CP), during which visual experience permanently molds many aspects of visual function (Hubel & Wiesel, 1970), spans the first through the fourth month in cats and about the 4.5th month through the third year in humans (Banks et. al., 1975). The receptive field properties which are susceptible to changes include acuity, orientation tuning, directional tuning and ocular dominance (cf. review by Movshon & Van Sluyters, 1981). For cats, the receptive field properties are most modifiable by visual deprivation at the onset of the CP, which is around the end of the first postnatal month (Hubel & Wiesel, 1970). Under normal rearing conditions, this sensitivity subsequently declines towards a minimal level at around the fourth month. In contrast, results of recent experiments suggest that the decline in visual cortical plasticity beyond the first month can be prevented for as long as neuronal activity is suppressed by dark rearing (Cynader & Mitchell, 1980; Mower et al, 1981). Further, once triggered for the onset of the CP by light, plasticity declines with a predetermined time course, irrespective of the subsequent visual environment (Mower et al, 1983). Thus, the dark rearing procedure offers a useful paradigm in which the progression of events associated with the decline of plasticity during the CP can be temporally separated from the progression of the other events that are associated with the general growth of the animal.

For this reason, we have collected samples of visual cortices from the following three groups of cats: (1) normal reared cats (NR) of ages 3-days to adults, so as to be able to characterize the normal ontogeny of biochemical events; (2) cats dark reared for 1-5 months (DR), so as to be able to identify the biochemical events that are postponed as the onset of the CP is postponed; (3) cats dark reared, then exposed to a few hours of light (DR + Lt), so as to be able to identify the biochemical events that first occur as the CP is triggered for its onset.

Realizing that many events could be responsible for determining the state of cortical plasticity, the study was begun by examining the role of hormone-stimulated cyclic AMP-metabolism in visual cortical plasticity. The reason for this was two-fold: 1) cyclic AMP has been implicated to control many developmental events in in vitro systems, such as the induction of the presumptive amphibian epidermis into derivatives of neural ectoderm and neural crest, and the stimulation of neurite outgrowth in neuroblastoma cell lines (Nathanson, 1977; 2) it has been hypothesized by Kasamatsu and his colleagues that a rise in intracellar cyclic AMP consequent to the release of norepineprine is a necessary event for the maintenance of visual cortical plasticity in cats (reviewed in Kasamatsu, 1985). The cyclic AMP-metabolism involves a cascade of events (figure 1), beginning with the binding of a hormone onto a receptor, followed by the stimulation of adenylate cyclase that is coupled to the receptor, causing a rise in the intracellular concentration of cyclic AMP. Cyclic AMP can then bind to a cyclic AMP-dependent protein kinase (cAMPdPK) that leads to an enhanced phosphorylation of various proteins. Alternatively, cyclic AMP can be hydrolyzed by cyclic-nucleotide phosphodiesterase (PDE). We have examined every step of this cascade within specific brain areas of NR, DR and DR + Lt animals, and the results are summarized below.

CYCLIC AMP-DEPENDENT PHOSPHORYLATION OF PROTEINS.

Current evidence indicates that in eukaryotes, every biological function modulated by cyclic AMP involves the stimulation of cAMPdPK (Nestler et al, 1984). Thus, the cyclic AMPdPK can be viewed as the final effector system within the cyclic AMP cascade of events. The in vitro phosphorylatability of proteins in whole homogenates of the visual cortex was assayed with and without cAMP and an exogenous supply of bovine heart cyclic AMPdPK.

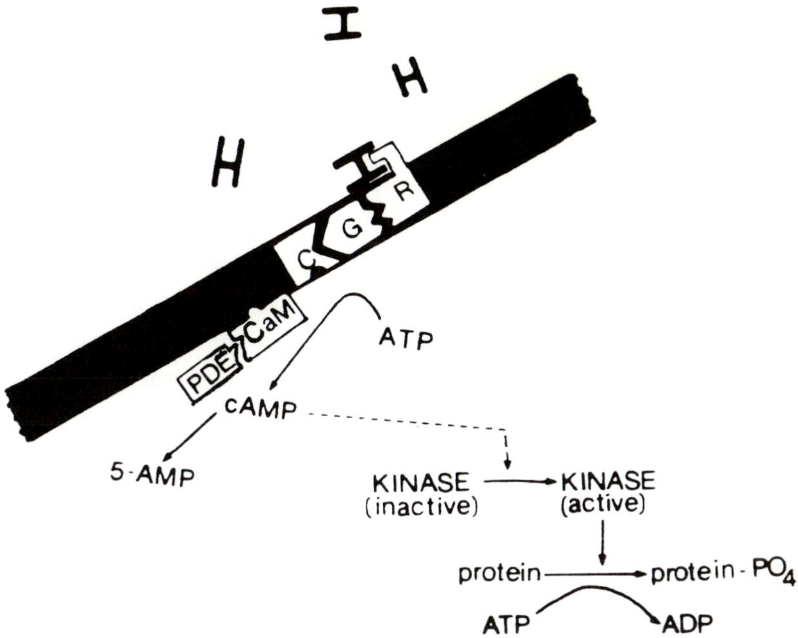

Figure 1. The cyclic AMP cascade of events. See text for details. Abbreviations are: H = hormone; R, G and C = receptor, GTP-binding, and catalytic subunits of the enzyme, adenylate cyclase; PDE = cyclic-nucleotide phosphodiesterase; CaM = calmodulin; KINASE = cyclic AMP-dependent protein kinase.

Following a 1 min incubation of homogenates with gamma -^{32}P-ATP, the phosphorylated proteins were separated by SDS-polyacrylamide gel electrophoresis, and counted for ^{32}P incorporation (Aoki & Siekevitz, 1985). When the tissues of DR and NR cats were compared, the in vitro phosphorylatability of one protein species was found to exhibit a strong dependence on the animal's rearing condition. This protein was immunologically identified (Aoki & Siekevitz, 1985) as microtubule-associated protein 2 (MAP2) that is found only in dendrites of neurons (Bernhardt & Matus, 1984). Specifically, following 2 months of DR from the time of eye-opening, its in vitro phosphorylatability was diminished, while the samples from the DR + Lt animals exhibited degrees of phosphorylatability that were equivalent to those of the NR cats (table I). The enhanced phosphorylatability of this protein species was seen among DR + Lt cats from three litters DR for 2 months, two litters DR for 3 months and one litter DR for 5 months. (Aoki & Siekevitz, 1985). Might the enhanced in vitro phosphorylability in the DR + Lt animals simply reflect the effect of dark-adaptation followed by exposures to light, or does this effect reflect a change underlying the onset of the CP? This question was addressed by examining the phosphorylatability of MAP2 in the visual cortex of animals that had been NR beyond the CP (7 months) to allow for normal development of visual function, then DR for 1 month, followed by re-exposure to light. The results (table II) indicated no change in the MAP2 phosphorylatability, suggesting that we had been visualizing an event that occurred with the onset of the CP. In the lateral geniculate nucleus (LGN), DR + Lt evokes neuronal activity, but will not trigger an onset of the CP, since the LGN has no postnatal CP to begin with (Movshon & Van Sluyters, 1981). Results (table III) indicated no change in MAP2 phosphorylatability here, supporting the idea that the enhanced MAP2 phosphorylatability reflects an event occurring at the onset of the CP. The auditory cortices of the same animals also did not exhibit any dependence on the environmental light (table IV, figure 2), confirming the specificity of the effect of sensory deprivation.

Since neither the MAP2 content nor the cAMPdPK content are influenced by the rearing condition (Aoki & Siekevitz, 1985), the change in the in vitro phosphorylatability implicates a change in the in vivo phosphorylation state of MAP2. This notion is corroborated by the observation that the cyclic AMP-stimulatable phosphorylation of the other major phosphoprotein in the brain, synapsin, is not influenced by light (tables I & IV): this finding also indicates that besides the cAMPdPK, both the ATPase and the phosphatase (if they act generally onto all phosphoproteins) also are probably not affected by the rearing condition. Specifically, the data suggest that the in situ MAP2 is relatively more phosphorylated under the DR condition, and becomes dephosphorylated with the onset of the CP. Since MAP2, when phosphorylated, induces microtubule depolymerization (Jameson et al, 1980; Jameson & Caplow, 1981; Murthy & Flavin, 1983) and the relaxation of the cross-linking of cytoskeletal proteins in dendrites (Nishida et al, 1981; Selden & Pollard, 1983) the dendritic shape would be expected to remain more malleable under the DR condition. Taken together, results suggest that visual cortical plasticity may be maintained at a high level by

TABLE 1

THE EFFECT OF LIGHT-EXPOSURE FOLLOWING DARK REARING ON THE RELATIVE PHOSPHORYLATION OF MAP2 AND SYNAPSIN IN THE VISUAL CORTEX HOMOGENATES OF CATS OF DIFFERENT AGES.

Tissue Source	MAP2	Synapsin	MAP2/Synapsin
52d DR	1.00 (153 cpm)	1.00 (855 cpm)	1.00
52d DR + 6hr Lt	3.48	1.09	3.19
52d DR + 6hr Lt	3.38	0.89	3.80
58d NR	3.58	-	-
58d DR	1.00 (381 cpm)	1.00 (837 cpm)	1.00
58d DR + 3hr Lt	1.36	0.74	1.84
58d DR + 7hr Lt	1.43	0.81	1.76
58d NR	1.44	-	-
98d DR	1.00 (257 cpm)	1.00 (696 cpm)	1.00
98d DR + 7hr Lt	2.18	1.07	2.04
97d NR	1.11	0.82	1.35

(from Aoki & Siekevitz, 1985)

the DR procedure, at leasts in part, by the maintainance of a malleable cell shape, which is in turn dependent upon the maintenance of MAP2 in a highly phosphorylated state. The consequence of the malleable cell shape may be that the visual cortical microcircuitry may be allowed to become rewired in response to biased visual experiences (=biased neuronal activity); conversely, plasticity may begin to decline following the exposure of animals to light, partly due to the dephosphorylation of MAP2 that results in an increased rigidity of cell shape. How might the in vivo state of phosphorylation of MAP2 be altered by the DR procedure? The results summarized above suggest that the changes may occur at some steps within the cyclic AMP cascade of events that precedes the activation of the kinase system.

ADENYLATE CYCLASE SYSTEM

We have examined the specific activities of the enzyme, adenylate cyclase (ACase), in homogenated tissues from the three groups of animals. This involved determining the basal activity, the GTP-stimulated activity, the GTP+hormone-stimulated activity

TABLE II

THE EFFECT OF DARK REARING (DR) IN ADULTHOOD ON THE RELATIVE PHOSPHORYLATION IN THE VISUAL CORTEX OF MAP2 AND SYNAPSIN

I. Tissue Source	MAP2	Synapsin	MAP2/synapsin
Adult, 2 mo DR	1.00 (661 cpm)	1.00(752 cpm)	1.00
Adult, 5 mo DR	1.64	1.27	1.29
Adult, never DR	1.66	1.85	0.90
Adult, never DR	1.39	1.03	1.35
II			
213dNR+30dDR	1.00 (379cpm)	1.00 (1040cpm)	1.00
213dNR+30dDR +3.5 hr Lt	0.84	1.00	1.84
213dNR+30dDR + 4 hr Lt	1.10	0.96	1.15

(from Aoki & Siekevitz, 1985)

(here, it was norepinephrine and dopamine) and the maximal catalytic activity in the presence of 10 mM of Mn^{++} and 140 μM of GppNHp. Among the NR cats, there was observed a concerted 10-fold increase of the basal, the norepinephrine-stimulated and the maximal catalytic activities during the period from postnatal day 3 up to 1-month of age, and a plateau in its activities beyond (Aoki, 1985a). The results also indicated that while the basal activities were not dependent upon the rearing condition at any age, the norepinephrine stimulatable ACase activity was: specifically, the norepineprine-stimulatable ACase was enhanced among the tissues obtained from DR and the DR + Lt animals more than that in the age-matched NR animals (Aoki, 1985a). This enhancement could be accounted for by the enhanced GTP-stimulability of ACase. This result, in turn, indicates that the coupling of the receptor subunit to the catalytic subunit that is controlled by the GTP-binding subunit and GTP (Schramm & Selinger, 1984) is increased in its efficiency (Aoki, 1985a). Thus, if the norepinephrine content within the visual cortex of DR cats is comparable to the content in NR cats, then it may be that MAP2 remains more phosphorylated due to the higher concentration of cyclic AMP resulting from the enhanced activity of norepinephrine-stimulatable ACase.

TABLE III

THE EFFECT OF LIGHT EXPOSURE FOLLOWING DARK REARING ON THE RELATIVE PHOSPHORYLATION IN THE LATERAL GENICULATE NUCLEUS OF MAP2 AND SYNAPSIN

Tissue Source	MAP2	Synapsin	MAP2/Synapsin
58d DR	1.00 (263 cpm)	1.00 (562 cpm)	1.00
58d DR + 3hr Lt	0.93	1.34	0.69
58d DR + 7hr Lt	1.00	1.20	0.83
58d NR	1.00	1.38	0.72
52d DR	1.00(211 cpm)	1.00(681 cpm)	1.00
52d DR + 6hr Lt	1.03	0.86	1.20

(from Aoki & Siekevitz, 1985)

CYCLIC NUCLEOTIDE PHOSPHODIESTERASE ACTIVITY (PDE).

The postnatal ontogeny of PDE is more modest compared to that of ACase: there is an approximately two-fold increase in its specific activity from postnatal day 3 up to 1 month of age, and a plateau in its activity beyond (Aoki, 1985a). This enzyme's specific activity was enhanced among DR + Lt animals compared to the DR littermates, while the values in tissues of the NR and the DR animals were not significantly different (Aoki, 1985a). Combining these data with that of the ACase, we can conclude that the concentration of cylic AMP within the cells of the visual cortex of DR animals remains relatively higher than the level in NR animals: once light evokes neuronal activities, the rate of turnover of cyclic AMP (including the rate of its breakdown) accelerates. The increased intracellular level of cyclic AMP may induce a stronger activation of cAMPdPK. On the otherhand, how the rate of turnover of cyclic AMP relates or to its ability to activate cAMPdPK is not known.

TABLE IV

A COMPARISON BETWEEN VISUAL CORTEX AND AUDITORY CORTEX OF THE EFFECT OF LIGHT EXPOSURE FOLLOWING DARK REARING ON THE RELATIVE PHOSPHORYLATION OF MAP2 AND SYNAPSIN

Tissue Source	MAP2	Synapsin	MAP2/synapsin
V-52d DR	1.00(838 cpm)	1.00(3160 cpm)	1.00
V-52d DR + 6hr Lt	2.55	0.92	2.78
A-52d DR	1.00(1971 cpm)	1.00(2993 cpm)	1.00
A-52d DR + 6hr Lt	0.82	0.78	1.04
A-52d DR + 6hr Lt	1.50	1.40	1.07
A-59d DR	1.00 (697 cpm)	1.00(1144 cpm)	1.00
A-59d DR + 6hr Lt	0.99	1.03	0.97

(from Aoki & Siekevitz, 1985)

THE LAMINAR DISTRIBUTION PATTERN OF RECEPTORS COUPLED TO ADENYLATE CYCLASE.

As already noted above, MAP2 is exclusively localized within dendrites. However, beyond this knowledge, we do not know where the changes in the state of phosphorylation of MAP2 may occur. An attempt to localize the site of this event was made by examining the laminar distribution pattern of the receptors known to be coupled to ACase. For this purpose, brain sections mounted on slides were incubated with radioactive ligands so that the ligand-binding sites (=receptor sites) could be visualized autoradiographically. The receptors examined so far are the beta-1 and beta-2 adrenergic receptors (Aoki et al, 1985) and the adenosine Al receptors (Aoki, 1985). As shown in figure 3, the laminar distribution in adulthood is strikingly similar for all of the receptors, having the highest density within laminae I through III, much less in lamina IV, and reappearing in laminae V and VI. However, the ontogenetic time courses for these receptors are different: while the adenosine Al receptor system becomes adult-like at around the third postnatal month, the beta-1 and beta-2 adrenergic receptor systems appear adult-like by the first postnatal month, or the beginning of the CP. Upon DR, none of these receptors exhibited any perturbation detectable in their ontogeny of the laminar distribution pattern. Thus, the changes appear not to occur at the level of the spatial

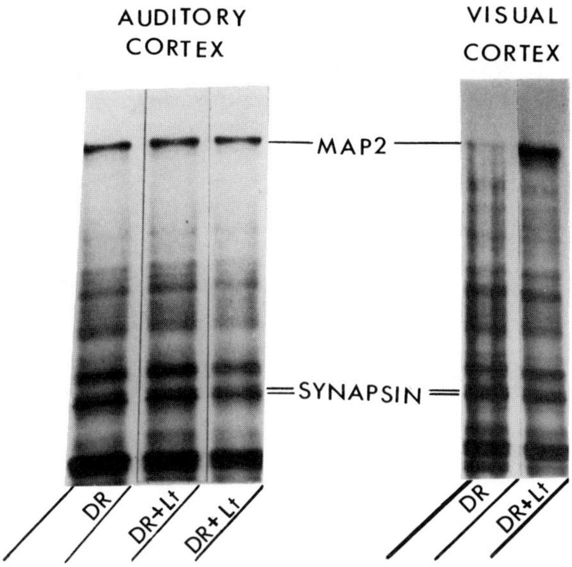

Figure 2. A comparison of the effect of exposure to light following dark rearing on phosphorylation of MAP2 in the auditory cortex and visual cortex.

All samples were whole homogenates from a 52d-DR litter. Whole homogenates containing 50μg protein were phosphorylated in the presence of gamma-^{32}PATP, 1.0 μM ccyclic AMP and exogenous bovine cyclic AMP-dependent protein kinase. Samples were then electrophoresed through SDS-polyacrylamide gels, so as to separate the protein species by size. Shown here is the autoradiogram of the gel in which the proteins that were phosphorylated in vitro can be identified as radioactive bands. (From Aoki & Siekevitz, 1985)

distribution pattern of these receptors. There are other receptors known to be coupled to the enzyme, ACase, which would be interesting to examine in the future.

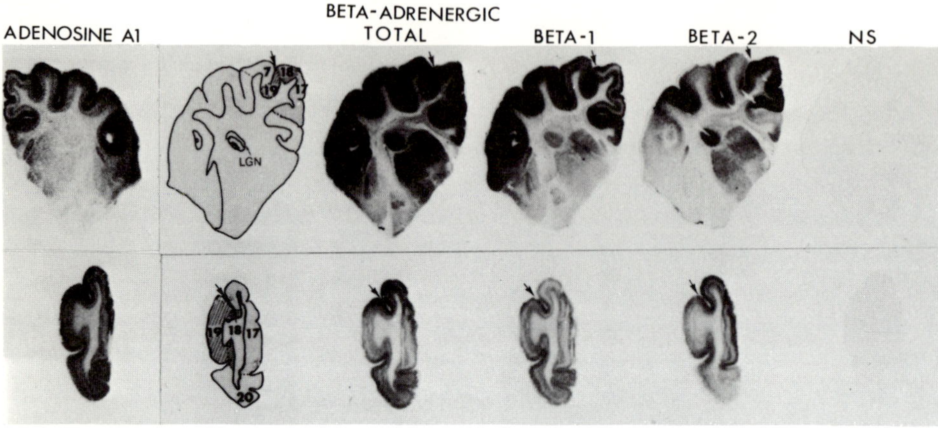

Figure 3. Comparison of the laminar distribution patterns of adenosine A1, beta-1 adrenergic and beta-2 adrenergic receptor sites.

Coronal sections from adult brains were incubated with radioactive ligands for the identification of the sites of the three receptors. Shown here are the autoradiograms obtained from exposing such sections to LKB Ultrofilms. The nonspecific (NS) labeling was undetectable in every experiment, indicating that the radioactivity reflected specific binding. Total = labeling of beta-1 + beta-2 + nonspecific binding sites.

SUMMARY

Upon examination of the cAMP cascade of events, it was observed that the state of the in vivo phosphorylation of MAP2 is enhanced in visual cortices that are plastic. When the decline in plasticity is triggered by light, MAP2 becomes dephosphorylated. Thus, visual cortical plasticity may depend, in part, on the cytoskeletal structure within dendrites that is maintained relatively malleable, due to the phosphorylation of MAP2. This, in turn, may allow for the restructuring of the visual cortical microcircuity in response to biased visual experience. This change in the state of phosphorylation may be due to the change in the chronic intracellular concentration of cyclic AMP or the rate of turnover of cyclic AMP. The sites of action of this event is most likely confined to dendrites, since MAP2 is found exclusively in dendrites. Further, this event may be more active within lamina I through III than in other lamina, since this is where the beta-1 and beta-2 adrenergic and adenosine A1 receptors are most concentrated.

REFERENCES

Aoki, C. (1985). Development of the A1 adenosine receptors in the visual cortex of cats, dark reared and normally reared, Dev. Brain Res., in press.

Aoki, C. (1985a). The role of hormone-stimulated cAMP metabolism in visual cortical plasticity. Thesis, The Rockefeller University.

Aoki, C., D. Kaufman and T. C. Rainbow (1985). The ontogeny of the laminar distribution of beta-adrenergic receptors in the visual cortex of cats, normally reared and visually deprived. Dev. Brain Res. submitted.

Aoki, C. & P. Siekevitz (1985). Ontogenetic changes in the cAMP-stimulatable phosphorylation of cat visual cortex proteins, particularly of microtubule-associated protein 2 (MAP2): Effects of normal and dark rearing and of the exposure to light. J. Neurosci., in press.

Banks, M. S., R. N. Aslin and R. D. Letson (1975). Sensitive period for the development of human binocular vision. Science 190: 675-678.

Bernhardt, R. and A. Matus (1984). Light and electron microscopic studies of the distribution of microtubule-associated protein 2 in cytoskeletons. J. Comp. Neurol. 226: 203-221.

Cynader, M. and D. E. Mitchell (1980). Prolonged sensitivity to monocular deprivation in dark-reared cats. J. Neurophysiol. 43: 1026-1040.

Hubel, D. H. & T. N. Wiesel (1970). The period of susceptibility to the physiological effects of unilateral eye closure in kittens. J. Physiol. (London) 206: 419-436.

Jameson, L. and M. Caplow (1981). Modification of microtubule steady-state dynamics by phosphorylation of microtubule-associated proteins. Proc. Nat. Acad. Sci. (USA) 78: 3413-3417.

Kasamatsu, T. (1985). Norepinephrine hypothesis for visual cortical plasticity: Thesis, antithesis, and recent development. In R. K. Hunt, A. A. Moscona & A. Monroy (Eds.), Current Topics in Developmental Biology, Academic Press, New York, in press.

Movshon, J.A. and R.C. Van Sluyters (1981). Visual neural development. Ann. Rev. Psychol. 32: 477-522.

Mower, G.D., D. Berry, J.L. Burchfiel and F.H. Duffy (1981). Comparison of the effects of dark-rearing and binocular suture on development and plasticity of cat visual cortex. Brain Res. 220: 225-267.

Mower, G.D., W.D. Christen and C.J. Caplan (1983). Very brief visual experience eliminates plasticity in the cat visual cortex. Science 221: 178-180

Murthy, A.S.N. and M. Flavin (1983). Microtubule assembly using the microtubule-associated protein MAP2 prepared in defined states of phosphorylation with protein kinase and phosphatase. Eur. J. Biochem. 137: 37-46

Nathanson, J.A. (1977). Cyclic nucleotides and nervous system function, Physiological Reviews 57: 157-256

Nestler, E.J., S.I. Walaas & P. Greengard (1984) Neuronal phosphoproteins: Physiological and clinical implications. Science 225: 1357-1364.

Nishida, E., T. Kuwaki and H. Sakai (1981). Phosphorylation of microtubule-associated proteins (MAPs) and pH of the medium control interaction between MAPs and actin filaments. J. Biochem. 90: 575-578.

Schramm, M. & Z. Selinger (1984) Message transmission: receptor-controlled adenylate cyclase system. Science 225: 1350-1356.
Seldon, S.C. and T.D. Pollard (1983). Phosphorylation of microtubule-associated proteins regulates their interaction with actin filaments. J. Biol. Chem. 258: 7064-7071.

Development of Stereopsis in Infants

Bela Julesz

For developmental psychology visual deficiencies are of great interest, particularly if they are not entirely the result of inherited genetic anomalies, such as color deficiences. Indeed, visual deficiencies after birth, during the critical period of plasticity, can teach us how the proper visual environment performs "fine tuning" of the nervous system for correct performance. Of course, in many cases visual deprivation is caused by genetic defects whose manifestations could be easily corrected by physical means if such deprivation had been noticed before permanent damage occured. Such a typical example is the permanent loss of neural pools tuned to specific orientations caused by severe astigmatism. For instance, Donald Mitchell (1980) reported that one identical twin whose astigmatism was corrected at 5 years of age had no neural deterioration, while his twin brother, whose astigmatism of similar severity was corrected only at the age of 7 years, had considerable neural loss of orientation. If this observation is corroborated, then the critical period for neural orientation tuning is amazingly late, somewhere between 5 and 7 years of age.

Stereodeficiences (including total loss of stereopsis, to be called stereoblindness) seem particularly important for developmental psychology. First, they affect, in my judgment (Julesz, 1971), about 15% of the human population, out of which 2% are stereo blind. Second, the manifestations often are quite dramatic resulting in a "lazy eye" condition. Third, the critical period of this loss of functional binocularity appears to occur rather early, although its exact onset is still not established. The thrust of my research interest in this area is to find ways to determine this critical period in infant monkeys and infant humans.

Here I will review some methodological innovations by my collaborators and me, and others, that permit the early diagnosis of functional binocularity in infants and some of their implications. This material has been published in the last 4 years, and the specialist familiar with these findings might wonder why I bother with a review talk. My only excuse is that during the same four years I have made discoveries with coworkers that led to a texton theory of preattentive vision, and in the light of these discoveries stereoblindness becomes one of the "texton blindnesses". This generalization raises some strategic questions for the development of preattentive vision.

Before going into these questions I give a brief account of two topics: A) the development of stereopsis in infants, and B) the texton theory of human vision.

THE DEVELOPMENT OF STEREOPSIS IN HUMAN INFANTS

There are several review articles on binocular depth perception in normal human adults and infants such as Julesz (1983), Julesz & Kropfl (1982), and Fox (1981). The interested reader should consult these reviews. Here, I give only a brief outline of the most important steps that led to a robust test of stereopsis in human infants.

Binocular depth perception is mediated by several mental processes, out of which stereoscopic depth perception, or briefly stereopsis, is by far the most important. Stereopsis yields the percept of relative depth, based on binocular disparity (briefly disparity) between corresponding points in the left and right eye's view. One can speculate why stereopsis has evolved in many animals, including monkeys and humans. In my opinion, it evolved to break camouflage, as I demonstrated in 1960 with the introduction into psychology of computer-generated random-dot stereograms (RDS) and random-dot correlograms (RDC). Some authors recently call these random-element stereograms, but as their inventor I retain the random-dot stereogram nomenclature, since from dots one can build up any local element. Such a stereogram is shown in figure 1, where the left and right arrays are composed of random black and white dots, and appear as random textures when viewed separately, but certain binocularly correlated areas jump out in vivid depth when binocularly fused. Binocular fusion of figure 1 can be easily

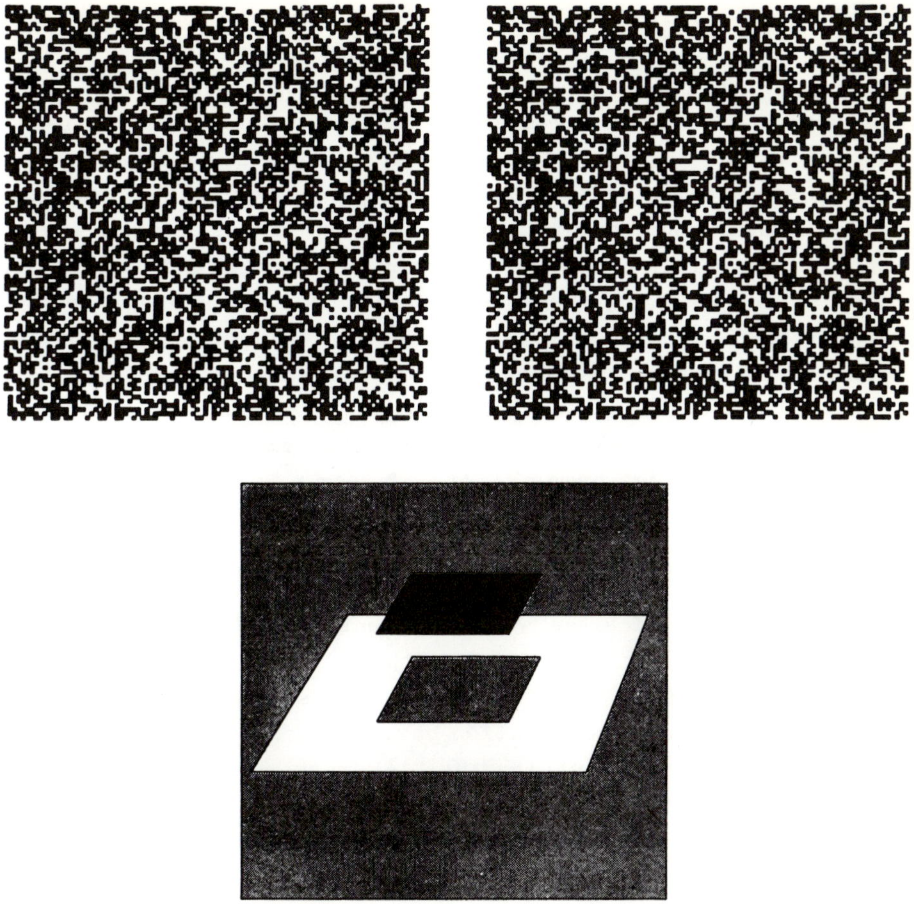

Figure 1. Static random-dot stereogram. When monocularly viewed no global shape or contour can be seen, but when stereoscopically fused a center square appears in vivid depth above the background. (Julesz, 1971).

accomplished by some, but requires optical aid by others. The easiest way to fuse these images is when they are presented as a red-green anaglyph, and are viewed through red-green goggles (such as provided in Julesz, 1971; 1978). Since these stimuli contain the correlated information only at a central site, the "cyclopean retina", I often refer to RDS and RDC as cyclopean stimuli.

In recent years, as computers became faster and acquired more memories, it has become possible to generate dynamic RDS of TV quality at 30-60 frames/s rates, although such dynamic RDS have been generated as computer-movies a decade earlier (Julesz, 1971). At the moment it is possible to generate a 250 x 250 dot dynamic RDS on a personal computer with 32 bit microprocessor at 30 frames/sec rates. Such dynamic RDS appear as dynamic noise when viewed monocularly, but do portray areas hovering in depth and also moving or pulsating in depth when viewed binocularly. With the help of dynamic RDS, and an even simpler class of stimuli, the so-called dynamic RDC (where the dynamic noise presented to either eye is identical for some period, and becomes uncorrelated for the next period) we could measure some basic spatiotemporal properties of stereopsis in normal human adults (Julesz & Tyler, 1978; Julesz, Breitmeyer & Kropfl, 1976; Tyler and Julesz, 1980).

Since dynamic RDS do not contain any monocular depth and familiarity cues, and at 60 frames/sec rates the monocular contrast at the boundaries of areas with different binocular disparities is practically zero, one could conclude that the only clue remaining is binocular correlation. That neural units must exist that are tuned to extract binocular correlation having specific disparities was psychophysically demonstrated by Blakemore and Julesz (1971) by producing depth after effects with dynamic RDS. Only recently did Gian Poggio (1984) find cyclopean neurons (that fire for dynamic RDS tuned to specific disparities) in Area 17 layer IVB and IVC of the monkey cortex. He also found units that fired for dynamic RDC as binocular correlation suddenly changed.

For years psychophysical studies with RDS on normal human adults yielded a vast amount of data and deep insight into the mechanism of stereopsis, and the interested reader might consult a review chapter by Julesz and Schumer (1981). However, psychophysical studies are diffucult with uncooperative observers, such as animals or human infants. So, in the next paragraph I turn to methods that use cortically evoked potentials to dynamic RDS or RDC.

The first cortically evoked potential that was recorded by scalp electrodes to study stereopsis was by Regan and Spekreijse (1970) who used static RDS, by switching from a zero disparity to a disparity mode. However, it has been suggested (Julesz, 1971; Julesz, et al., 1980) that the observed potentials could have been contaminated by eye-movement potentials or by monocular cues present when rapid changes are made in static RDS or when monocular contours are present.

Such cues, of course, are totally absent in dynamic RDS. Bouldin et al. (1975) used dynamic RDS to study the origin of a cortically evoked potential, known as the lambda wave, which is believed to reflect the operation of a saccadic eye-movement suppression mechanism. Lambda waves can be elicited either by moving the eyes over a stationary contour or by moving a contour over immobilized eyes. Bouldin et al. generated cyclopean contours that moved across the screen while the observer's eyes were held immobile by fixation markers. Evoked potentials were elicited as long as the cyclopean contour produced good apparent movement. The latency of the potentials elicited by cyclopean moving contours was about 100msec slower than those elicited by classical luminance contours. Bouldin (1975) also observed that the magnitude of the potential varied with disparity.

In 1976, Lehmann & Julesz studied cyclopean targets pulsating in depth on the left or right hemifields and measured the evoked potentials with a multiple electrode array over the scalp. We wanted to know whether the evoked potential was merely the result of detection of the depth pulsation or was directly related to the site of stimulation (Lehmann and Julesz, 1978). We measured much larger EP by electrodes contralateral to the cyclopean stimulation site.

Rawlings and Yates (1978) generated dynamic RDS in anaglyph format on a color projection TV system and obtained potentials to cyclopean checkerboards pulsating in depth. We (Julesz et al., 1980) also used dynamic cyclopean checkerboards pulsating in depth, but in addition we used RDC that consisted of the same dynamic noise sequence in both eyes' views followed periodically by a complemented array of noise in one eye's view with respect to the other. In figure 2 typical responses are shown for two observers when depth at a cyclopean checkerboard (B) or correlation (A) fluctuates at 1 Hz rate (Julesz et al. 1980). Figure 2c is a monocular control (when one eye is closed), while figure 2d is similar to 2a but the anaglyph was viewed without the red-green goggles, giving rise to monocular cues. Comparison between figures 2a and 2b reveals great differences between evoked potentials (EP) to RDC and RDS. The shape of the former has an almost rectangular appearance as binocular fusion and rivalry alternate.

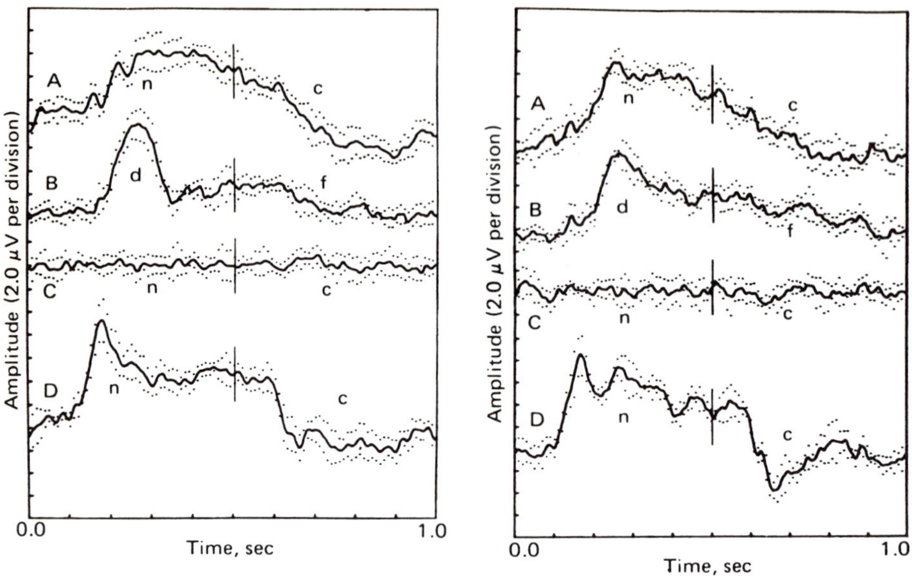

Figure 2. EP to two reversals per sec stimuli for subjects WK (right) and DB (left) belonging to stimulus conditions: (A) Dynamic RDC. Negative correlated (anticorrelated) interval denoted by n, correlated interval by c. (B) Dynamic RDS portraying a cyclopean checkerboard pulsating in depth from a flat background. The depth interval is denoted by d, the flat interval by f. (C) Monocular control. same as A but with one eye closed, while other eye saw through a red (or green) filter. (D) Same as A but anaglyph was viewed without redgreen goggles, giving rise to monocular color cues. Solid line is mean of 6 EP (whose N=50). Dotted lines are ±1 SD of the 6 EP. Positive polarity, measured on the anterior electrode, is plotted upward. (Julesz, et al, 1976).

We developed this robust technique, particularly when the cyclopean stimuli are projected on a large (6-foot) screen, to test the stereopsis of human infants in a brief period (of a few minutes). The large screen surrounds the observer with the stimulus, so it is not very critical whether uncooperative observers (such as infants or monkeys) move their eyes during the test.

The dynamic RDC are most useful for uncooperative subjects, since these stimuli have no disparity (correlation and negative correlation alternate without having to maintain horizontal head positions), while for RDS horizontal disparities are critical. However, it is not clear whether EP to RDC are measuring stereopsis or only functional binocularity of finding differences between responses to correlation or negative correlation. In figures 3a and 3b one can see EP to RDC and RDS, respectively at 3 Hz alternation rates for normal adults (Julesz, et al., 1980). When a cyclopean checkerboard jumps in front or recedes in the surround at 3 Hz rate, the EP wave has also a 3 Hz periodicity (figure 3b). However, for a RDC that alternates at 3 Hz rate, the corresponding EP wave has a 6 Hz periodicity. I will return to this problem later.

Fortified with these techniques we undertook a screening test on human infants at the Mount Sinai Hospital (Braddick, et al., 1980), and Benno Petrig, a former doctoral student of mine at ETH (Zurich) undertook another test on infants (Petrig, et al., 1981). We were able to show that functional binocularity in normal humans occurs at the age of 10 - 19 weeks, while EP to classical checherboards (where luminance varies) occurs much earlier (figures 4 and 5). From figure 5 one can observe the doubling of EP to RDC at 6 months of age, which suggests that functional binocularity might precede stereopsis, and EP shapes to RDS, indicating stereopsis, reach adult shapes at age 6 months, or somewhat before.

I emphasize the EP method elicited by cyclopean stimuli over preferential looking techniques for obvious reasons, particularly that the former is more robust and takes only a few minutes (after the electrodes are attached. Atkinson and Braddick (1976) used static RDS and preferential looking techniques on four 8-week old infants, and found only one infant showing evidence of stereopsis. Fox et al., (1980) used dynamic RDS and preferential looking techniques and found the median age of onset of stereopsis to be 127 days, or about 4 months. This is very similar to the Braddick et al. (1980) and Petrig et al. (1981) findings of 3 to 3-1/2 months. It should be noted that Petrig et al. (1981) also used cyclopean checkerboards besides RDC, so the somewhat earlier onsets might be due to the more sensitive method of EP versus preferential looking. That Braddick et al (1980) found four infants 2 months or somewhat younger showing EP to dynamic RDC might be the result of functional binocularity preceding global stereopsis. From all reported studies it is safe to assume that at 6 months of age global stereopsis (i.e. stereopsis of RDS) is fully developed in the majority of infants.

In all these studies the onset of global stereopsis has been measured by relatively large disparities of as much as 40 minutes of arc. This should not be confounded with the development of stereoacuity that for adults is a few seconds of arc. While I

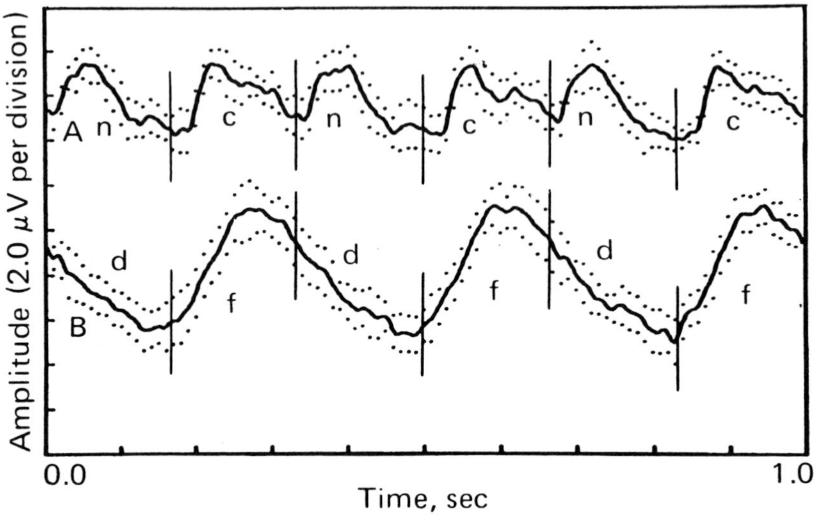

Figure 3. EP to six reversals per sec-stimuli for subject WK under stimulus conditions A and B, respectively, as described for figure 5. Solid line is mean of 12 EP (whose N=75) for a total of 900 individual responses. Dotted lines are ± sd of the 12 EP. (Julesz, et al, 1976).

restrict this review to global stereopsis and skip work with classical stereograms that contain monocular cues (e.g. the Titmus test), I briefly note the work of Romano et al. (1975). They used the Titmus test (which is limited to 40 sec arc minimum disparity), and found that only one child reached 40 sec arc stereothreshold at 34 month age, while in the age range 3 to 4-1/2 years, 13 of 87 (15%) achieved 40 sec. From this it seems that a very few infants might have respectable stereoacuities; it takes several years until it decreases to adult thresholds.

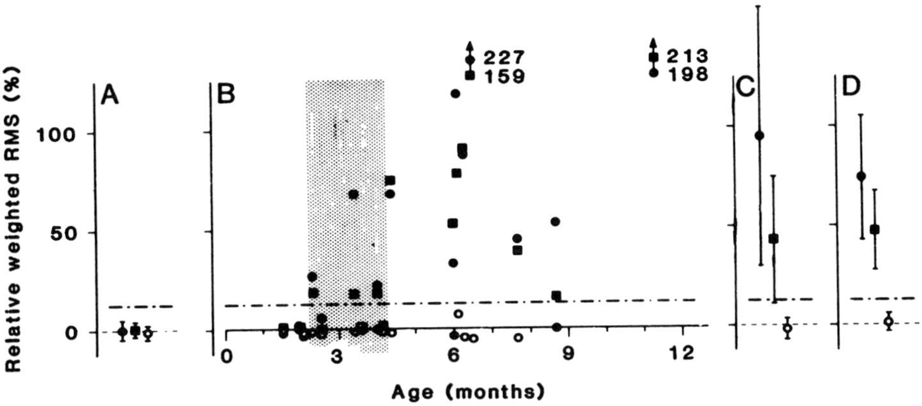

Figure 4. Relative weighted root-mean-square (RMS) Values of the two stereoscopic test stimuli NP ( ) and DF ( ) and of the control stimulus SS (0). NP: Random-dot correlogram, DF: Cyclopean checkerboard, and SS(0): Snowstorm with zero disparity. Data are given for (A) stereoblind adults and Children (N=15), (B) infants (N=17), (C) stereonormal children (N=20), and (D) stereonormal adults (N=20). Criterion responses (relative weighted RMS 12 percent) are above the dashed line. The shaded zone (2 to 4.3 months) is the age range where cyclopean VEP's developed in our sample of infants. Data in (A), (C) and (D) represent means ± 1 standard deviation. (Petrig et al. Copyright 1981 by the AAAS.)

Recently, we tried some RDS with sinusoidally corrugated depth surfaces (Schumer & Julesz, 1984) and determined the cyclopean modulation transfer functions (CMTF). Figure 6A shows the appearance of such a corrugated stimlus, and figure 6B shows the CMTF for different depth pedestals for normal human adults. It is surprising how low the cutoff threshold is (about 3 cycles/deg) for the spatial frequencies of corrugation. It would be interesting to study these CMTF for all ages.

TEXTONS OF PREATTENTIVE VISION

It took me and my collaborators 20 years to show that preattentive vision (particularly texture discrimination) is based on a few conspicuous elements, I named textons (Julesz, 1962, 1981, 1984; Caelli and Julesz, 1976; Bergen and Julesz, 1983; Julesz and Bergen, 1983). The textons are elongated blobs of certain colors, widths, length, orientations, depths, velocities, and flicker rates. For more elongated blobs (such as line segments) their crosssings

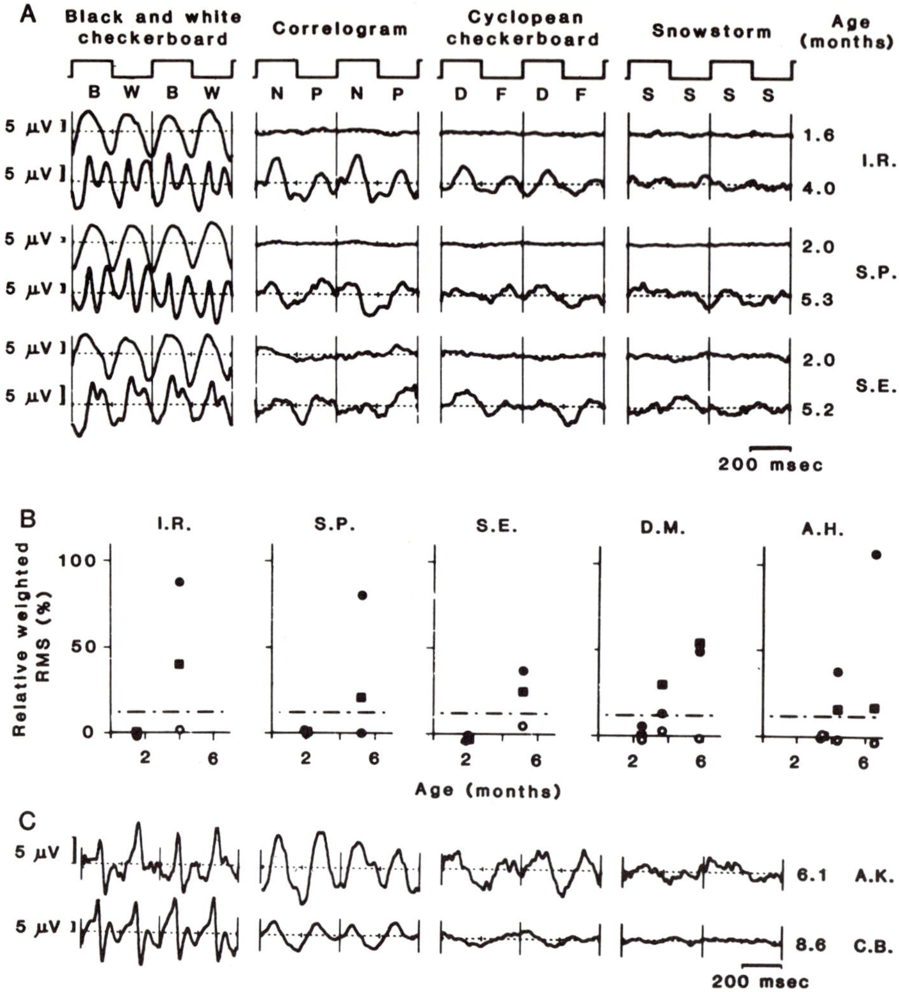

Figure 5. (A) The VEP's to the four stimuli recorded from three infants (I.R., S.P., and S.E.) at different ages. Dashed line represents computed zero after elimination of the direct-current voltage offset. Positivity at the occipital electrode results in a downward deflection of the trace. The VEP's for each subject are scaled so that the RMS value in the first column (stimulus BW) is unity. Each VEP shows two stimulus periods (delimited by the long vertical lines) each of which consists of two stimulus halves (marked by a short vertical line). (B) Relative weighted RMS values derived from the VEP's of five initially stereoblind infants retested at different ages (see figure 2). (C) Same as (A), but for subjects A. K. and C. B. Note doubling of VEP period to NP as opposed to DF. (Petrig, et al. Copyright 1981 by the AAAS.)

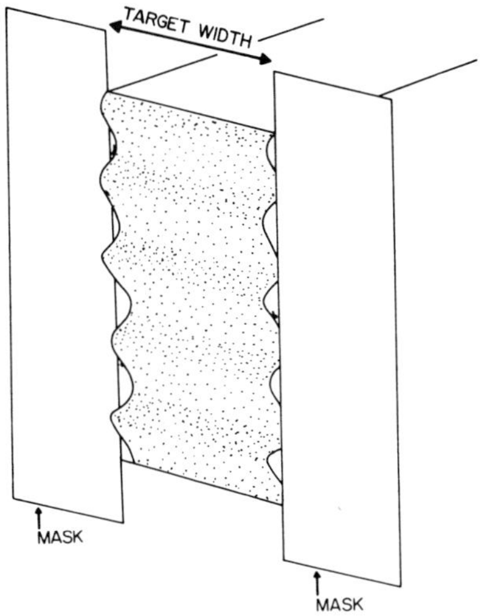

Figure 6A. The stimulus appears as a corrugated surface in depth, with a dotted texture. The lateral edges are masked in the actual display so that monocularly present edge ripples are not seen, and the spacing of possible dot positions is substantially smaller than the width of an actual dot. The stimulus is thus "cyclopean". (Schumer and Julesz, 1984).

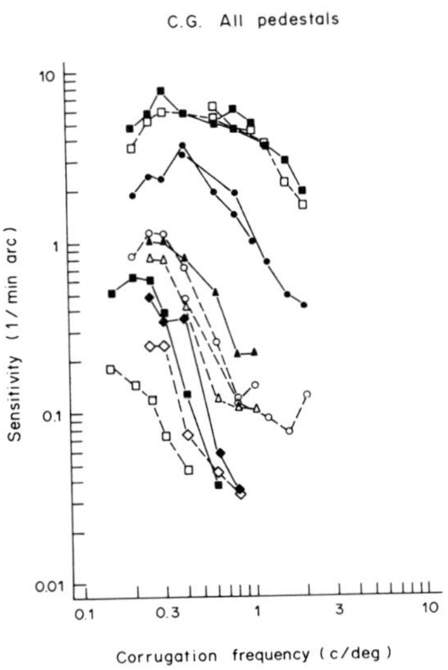

Figure 6B. Summary of data for C. G., plotted on one graph for comparison across pedestals. Squares: no pedestal. Circles: 20' arc pedestal. Triangles: 30' arc pedestal. Diamonds: 40' arc pedestal. Squares: 50' arc pedestal. solid symbols and solid lines: uncrossed pedestals. Open symbols and dashed lines: crossed pedestals. Note greater sensitivity to uncrossed pedestals than to crossed pedestals. (Schumer, R. and Julesz, 1984).

and terminators are also textons. Here I do not want to go into detail, exept to show a few examples and refer the reader to the references given above. For instance, in figure 7 one can effortlessly (preattentively) detect an area composed of "+"s among "L"s, while it takes time-consuming effort through element-by-element scrutiny to detect an area composed of "T"s among the "L"s. This latter operation is called focal attention.

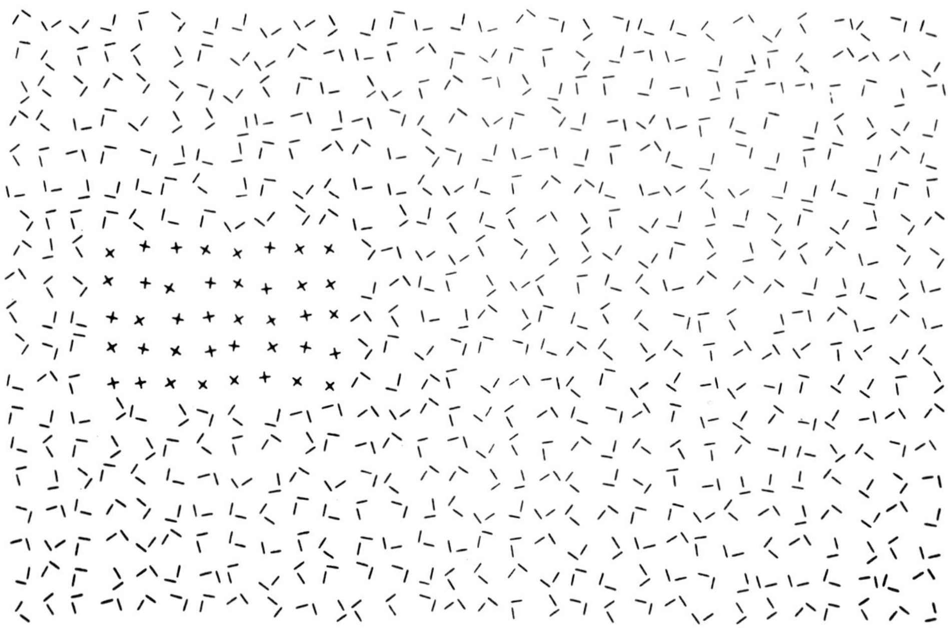

Figure 7. Preattentive texture discrimination (the immediate detection of '+'s in the 'L's versus focal attention (searching for the 'T's in 'L's by element-by -element scrutiny). (Julesz and Bergen, 1983).

The texton theory claims that without scrutiny - that is preattentively - one can detect only differences in textons or their densities (i.e. detect texton gradients), and the relative position between textons is ignored. However, if patterns or textures (i.e. aggregates of patterns), are composed of the same number of textons, they cannot be told apart without element-by-element scrutiny. Indeed, the "T" and "L"-shaped patterns consist of the same two perpendicular line-segments and preattentively cannot be told apart.

Perhaps two more demonstrations might give the flavor of the texton theory. Figures 8 and 9 show the importance of terminators (ends-of-lines). In figure 8 all line segments that make up the micropatterns of which the texture pairs are composed have the same length, width, orientation and color, but differ in their number of terminators (Julesz, 1981). This terminator difference (2 versus 5) yields strong preattentive texture discrimination. On the other hand, figure 9 shows the case where a pattern pair in isolation appears very different. However, it consists of identical textons and in the light of the texton theory should not yield preattentive discrimination. Indeed, one needs detailed element-by-element

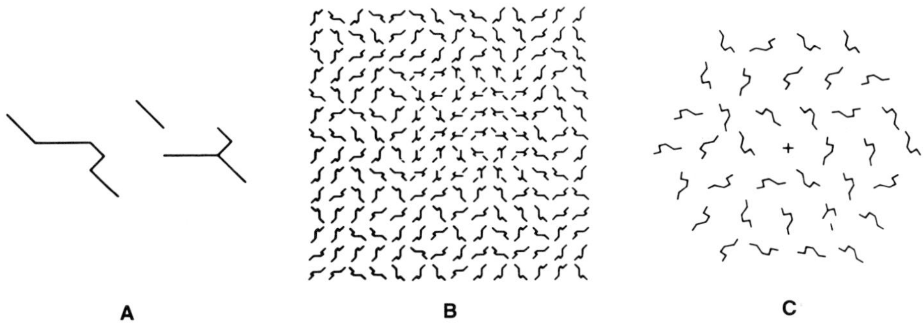

Figure 8. Preattentively discriminable texture pair (B) where the two elements shown in (A) have identical line-segment textons but differ in their number of ends of lines (terminators). (C) Preattentive discrimination is so strong that a single element can be detected among 35 dual elements in a 100-ms flash terminated by a masking stimulus. (Julesz and Bergen, 1983).

scrutiny to tell apart areas consisting of "S"-shaped elements from areas composed of "10" shaped patterns.

What does this preattentive texton theory imply for developmental psychology? I think a great deal. We know from the literature how orientation, contour and depth perception require some fine-tuning after birth. These are some of the textons, the fundamental local features (dimensions) of preattentive vision. I alluded to the fact that 4% of the population are color-deficient, 2% stereoblind, and some with severe astigmatism uncorrected in early years might have orientation-deficiencies. Here I suggest that terminator-deficient people might exist that might include some dyslexics.

In summary, if texton perception has been arrested in a critical period, then this loss of preattentive vision impairs figure-ground segregation, the primary purpose of preattentive vision. I think the study of visual development has to consider all the textons as a whole.

WHAT TO DO NEXT?

Now that a robust test exists for testing stereoblindness in infancy some important problems that might lead to their prevention can be raised. First of all, let me review some relevant studies of stereopsis in monkeys. Here I do not want to go into detail, but

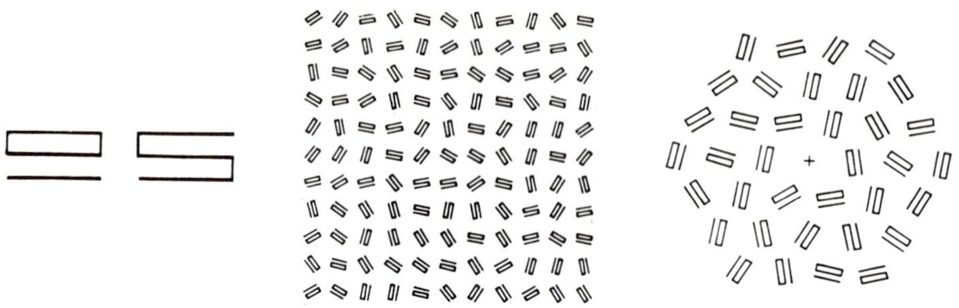

A B C

Figure 9. Preattentively indistiguishable texture pair (B) and indistinguishable target (C), although the elements themselves appear very different (A). (Julesz and Bergen, 1983).

only mention that the first behavioral study on macaque monkey with RDS was performed by Bough (1970) and on cat by Lemkuhle and Fox (1977). We were able to elicit EP to dynamic RDC in monkey during a prolonged stay of mine at Caltech (Miezin et al., 1981).

Hubel and Wiesel (1970) and Wiesel (1982) demonstrated the importance of artificially induced squint by unilateral eye closure on arresting the development of binocular neurons in the infant monkey cortex. This epoch making discovery has been followed by several interesting studies. Pettigrew and Kasamatsu (1978) showed that the period of plasticity in which ambloyopia ex anopsia could be introduced could be lengthened by supplying norepinephrine to the monkey's brain.

Recently, Crawford et al. (1984) raised infant monkeys with distorting prisms before their eyes for 30 days between 30-60 days of age. There followed a 3-year period of normal binocular visual experience. In spite of the fact that these early deprived animals had no strabismus they proved to be behaviorally stereoblind to RDS. After behavioral testing, the deprived animals had only 22% binocularly active neurons (of 880 neurons tested), while the control group (not deprived of binocular fusion in the critical period) had 81% functional binocular neurons. These experiments demonstrate: a) for the same subjects the strong association between binocular neurons and primate stereopsis; b) that congenital binocular neurons, once lost, do not recover even with extensive binocular experience; and c) emphasize the vulnerability of the primate binocular system to abnormal early visual experience.

From these findings on infant monkeys some plausible suggestions can be posed for human infants. First of all, while strabismus might sometimes be the effect of improper neural functioning of the stereoscopic system, in most cases it seems the reverse is the case. Strabismus is most likely the cause of neural defects of stereopsis. Whether the 1-2 months plasticity period in monkey is the same for human infants, or has to be multiplied by a factor of 2-3 to correspond to the onset of functional binocularity at 3-6 months, is a most important question to be answered. If the former is true, surgical correction of strabismus at 3-4 months of age might be too late. On the other hand, if the later onset of functional binocularity agrees with the critical period of plasiticity, then surgery at this later age is much preferred. One way to approach this problem is to measure cyclopean EP in infant monkeys, a project I am planning to do at Caltech.

Unfortunately, the relation between strabismus and permanent loss of stereopsis is complex. For instance, it is well documented in the ophthalmological literature that surgical correction of strabismus for esotropes (whose strabismus goes inward) yields much poorer rehabilitation of stereopsis than for exotropes (whose strabismus goes outward). I do not want to speculate why this is so, but if squint prevents the proper stimulation of corresponding points, then it would seem immaterial whether this was caused by extropic or esotropic strabismus. However, recent work by Tieman et al. (1983) and Tumosa et al. (1983), who produced amblyopia in cats by alternating monocular exposure found larger deficits in the nasal hemifields. Perhaps this imbalance between nasal and temporal hemifields to binocular deprivation might be related to the problems of esotropia causing more severe binocular dysfunction than exotropia.

The problem of how proper binocular stimulation affects the neural machinery of stereopsis is of interest to all of us. Usually as we perform vergence movements of our eyes, we bring areas of large disparities within Pamun's fusional area. So, most of us with good stereopsis are unaware of quite large scotomas to certain disparities that coexist in the same retinal positions with highly sensitive cyclopean units tuned to other disparities. Figure 9 shows a cyclopean perimetry study by us (Breitmeyer et al., 1976) using dynamic RDS with small 24 min arc x 24 min arc targets at $\pm$ 3 min arc disparities from the surround. We measured the duration threshold of seeing the target in msec, and as shown in figure 10 no left/right hemifield asymmetry exists for stereopsis, but the upper half field is more sensitive for uncrossed disparities, while the lower hemifield is more sensitive for crossed-disparities. These findings were anticipated by von Helmholtz who assumed that the horopter is tilted, and the observer stands on his own horopter. What is particularly interesting, however, is that for observer WK (upper two arrays), just 30 min arc below the center of fixation a "cyclopean scotoma" exixts for uncrossed disparities (177 and 174 msec). At the same locale for crossed-disparity (thus only 6 min arc away) the same observer has very sensitive stereopsis. Throughout many years as we tested observers with good stereopsis, we often noticed large cyclopean scotomas either for crossed or uncrossed disparities, that were revealed by brief exposure (preventing vergence movements).

Whether stereopsis could be improved by prolonged training for such "virgin areas" has been an interesting question and over many years, I used to present briefly RDS with large disparities (outside their Panum's fusional area) to summer students. After thousands of trials, many of these observers started to increase their Panum's area and were able to tell whether the cyclopean target was in front or below the surround, or was uncorrelated.

I mention these observations to point out the huge difference between unused stereodetectors and incorrectly used stereo detectors. The former did not functionally develop but seem to have the ability to become functional after proper and prolonged usage while the latter, caused by squint, or deprivation of detailed vision over one eye (e.g. a monocular cataract) became irreversibly arrested in their functioning.

Let me note that if the critical period for stereopsis were to occur in the first or second months in human infants, and this age would remain too early for surgical correction of strabismus, the RDS techniques could still prove useful. One could give strabistic infants dynamic RDS with slowly varying disparities so that in spite of their squint, some binocular disparity detectors might be properly stimulated. Perhaps for a subclass of infants an exercise of a few minutes per day may delay the onset of amblyopia to a safe period when the infants could be operated on.

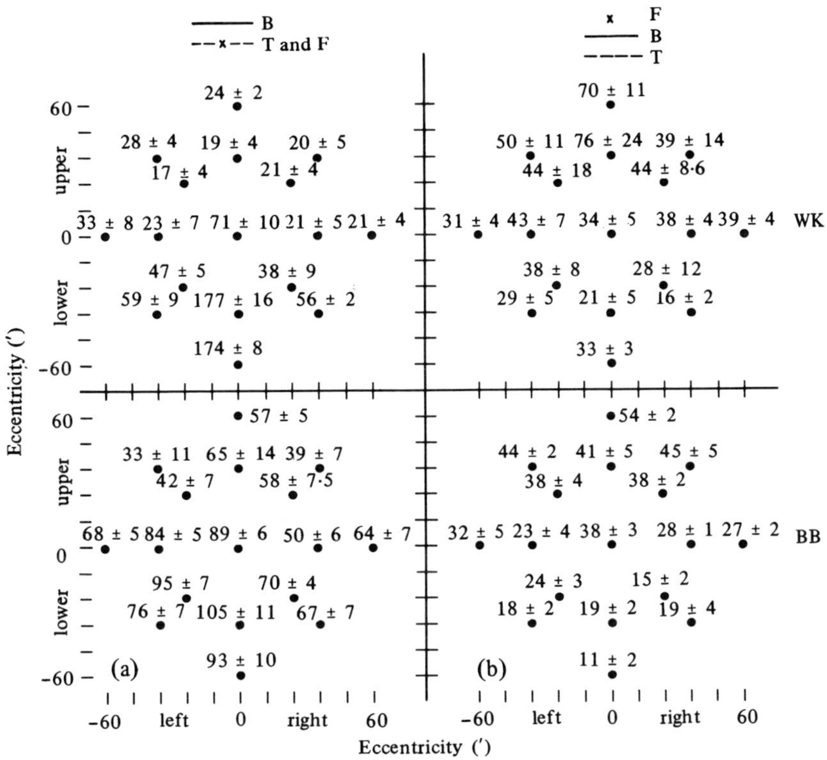

Figure 10. The means and standard errors (in ms) of duration thresholds (temporal limits) for detecting a 24' x 24' cyclopean test square centered at points as shown under (a) viewing condition 2 and (b) viewing condition 4. The fixation marker was located at the center of gaze (0,0). The abscissa and ordinate give horizontal and vertical visual field eccentricities and are graduated in units of 12'. (Julesz, Breitmeyer, and Kropfl, 1976).

Whether all the knowledge that slowly accumulates will lead to the prevention of stereoblindness in the foreseeable future remains to be seen, but it is certainly worth trying.

REFERENCES

Atkinson, J. and O. Braddick. 1976. Stereoscopic discrimination in infants. Perception 5: 29-38.

Bergen, J. R. and B. Julesz. 1983. Parallel versus serial processing in rapid pattern discrimination. Nature 303: 696-698.

Blakemore, C. and B. Julesz. 1971. Stereoscopic depth aftereffect produced without monocular cues. Science 171: 386-388.

Bough, E. W. 1970. Stereoscopic vision in the macaque monkey: a behavioral demonstration, Nature 225: 42-44.

Bouldin, D. W. 1975. Visual evoked cortical potentials elicited by dynamic random-dot stereograms. Doctoral dissertation, Vanderbilt University. Dissertation Abstracts International, 1976. 36, 3516B. (University Microfilms No. 76-92, 172).

Bouldin, D. W., J. R. Bourne, and R. Fox. 1975. Lambda waves elicited by cyclopean contour movement. Paper presented at the Association for Research in Vision and Ophthalmology, Sarasota, Forida, May 1975.

Braddick, O., J. Atkinson, B. Julesz, W. Kropfl, I. Bodis-Wollner, and E. Raab. 1980. Cortical binocularity in infants. Nature 288: 363-365.

Breitmeyer, B., B. Julesz, and W. Kropfl. 1975. Dynamic random-dot stereograms reveal an up-down anisotropy and left-right isotropy between cortical hemifields. Science 187: 269-270.

Crawford, M. L. J., E. L. Smith III, R. S. Harwerth, and G. K. von Noorden. 1984. Stereoblind monkeys have few binocular neurons. Invest. Ophthalmol. 25: 777-781.

Fox, R. 1981. Stereopsis in animals and human infants. In Development and Perception, Bol. 2, pp. 335-381. Academic Press.

Fox, R., R. N. Aslin, S. L. Shea, and S. T. Dumais. 1980. 207: 323-324.

Hubel, D. H. and T. N Wiesel. 1970. Cells sensitive to binocular depth in area 18 of the Macaque monkey cortex. Nature 255: 41-42.

Julesz, B. 1962. Visual Pattern discrimination. IRE Transactions on Information Theory IT 8: 84-92.

Julesz, B. 1971. Foundation of Cyclopean Perception (Univ. of Chicago Press: Chicago)

Julesz, B. 1978. Global stereopsis: Cooperative phenomena in stereoscopic depth perception. In: Handbook of Sensory Physiology, Vol. 8: Perception, eds. R. Held, H. W. Leibowitz, H. L. Teuber. Heidelberg: Springer-Verlag.

Julesz, B. 1981. Textons, the elements of texture perception, and their interactions. Nature 290: 91-97.

Julesz, B. 1984. A Brief outline of the texton theory of human vision. Trends in Neurosciences 7: 41.

Julesz, B., B. Breitmeyer, and W. Kropfl. 1976. Binocular-disparity-dependent upper-lower hemifield anisotropy and left-right hemifield isotropy as revealed by dynamic random-dot stereograms. Perception 5: 129-141.

Julesz, B. and C. W. Tyler. 1976. Neurontropy, and entropy-like measure of neural correlation in binocular fusion and rivalry. Biol. Cybern. 23: 25-32.

Julesz, B., W. Kropfl, and B. Petrig. 1980. Large evoked potentials to dynamic random-dot correlograms and stereograms permit quick determination of stereopsis. Proc. Natl. Acad. Sci. USA 77, No. 4: 2348-2351.

Julesz, B. and R. A. Schumer. 1981. Early visual perception. In: Ann. Rev. Psychol. 32: 575-627.

Julesz, B. and W. Kropfl. 1982. Binocular neurons and cyclopean visually evoked potentials in monkey and man. Annals of The New York Academy of Sciences: 37-44.

Julesz, B. and J. R. Bergen. 1983. Textons, the fundamental elements in preattentive vision and perception of textures. Bell Syst. Tech. J. 62, No. 6: July - Aug.: 1619-1645.

Lehmann, D. and B. Julesz. 1978. Lateralized cortical potentials evoked in humans by dynamic random-dot stereograms. Vision Res. 18: 1265-1271.

Lemkuhle, S. W. and R. Fox. 1977. Global stereopsis in the cat. Paper presented at the Association for Research in Vision and Ophthalmology, Sarasota, Florida, May 1977.

Miezin, F. M., J. Myerson, B. Julesz, J. M. Allman. 1981. Evoked potentials to dynamic random-dot correlograms in monkey and man: a test for cyclopean perception. Vision Res. 21: 177-179.

Mitchell, D. E. 1980. The influence of early visual experience on visual perception. In: Visual Coding and Adaptability, ed. C. S. Harris. Hillsdale, NJ: L. Erlbaum Associates.

Petrig, B., B. Julesz, W. Kropfl, G. Baumgartner, and M. Anliker. 1981. Development of stereopsis and cortical binocularity in human infants: electrophysiological eviodence. Science 213: 1402-1405.

Pettigrew, J. D. and T. Kasamatsu. 1978. Local perfusion of noradrenalin maintains visual cortical plasticity. Nature 271: 761-763.

Poggio, G. 1984. Processing stereoscopic information in primate visual cortex. In: Dynamic Aspects of Neocortical Function, eds. Edelman, cowan and Gall, New York: John Wiley & Sons.

Rawlings, S. C. and J. T. Yates. 1979. Visual evoked potentials from high-density dheckerboard depth-reversal stereograms. Paper presented at the Association for Research in Vision and Ophthalmology, Sarasota, Florida, May 1979.

Regan, D. and H. Spekreijse. 1970. Electrophysiological correlates of binocular depth perception in man. Nature 225: 92-94.

Romano, P. E., J. A. Romano, and J. E. Puklin. 1975. Stereoacuity development in children with normal binocular single vision. Amer. J. Ophthalmol. 79: 966-971.

Schumer, R. A. and B. Julesz. 1984. Binocular disparity modulation sensitivity to disparities offset from the plane of fixation. Vision Res. 24: 533-542.

Tieman, S. B. and H. V. B. Hirsch. 1983. Exp. Brain Res. 47: 119-129.

Tumosa, N.m, S. Nunberg, H. V. B. Hirsch, and S. B. Tieman. 1983. Binocular exposure causes suppression of the less experienced eye in cats previously reared with unequal alternating monocular exposure. Inv. Ophthal. 24/4: 496-506.

Wiesel, T. N. 1982. Postnatal development of the visual cortex and the influence of environment. Nature 299: 583.

Wiesel, T. N. and D. H. Hubel. 1965. Comparison fo the effects of unilateral and bilateral eye closure on cortical unit responses in Kittens. J. Neurophysiol. 28: 1029.

Redevelopment of Fixation and Scanning Eye Movements Following the Loss of Foveal Function

S.G. Whittaker and R.W. Cummings

A critical part of visual perceptions, our ability to orient to a visual environment and quickly identify its most interesting attributes, depends upon our control of eye movements. Within a few seconds of confronting a page of text or a city street our oculomotor control system organizes and executes a series of saccades. The eye darts about the scene, bringing aspects of interest into the narrow field of the fovea for the resolution of detail. In this manner, our oculomotor system paints a global perception of a visual scene or transmits a message from the printed page to the brain.

We suspect that the control of saccades is a phylogenetically ancient system, with the most primitive function being that of keeping the interesting aspects of a visual scene somewhat centered in the visual field. As an animal moves about, both the optokinetic (OKN) and vestibuloocular (VOR) systems counter head motion with eye motion, keeping the eye reasonably stable in space and the visual image reasonably stable on the retina. As the head, or a moving visual image forces the eye to an extreme position, a quick reset, the "quick phase", resets the eye to a highly predictable craniotopic position, 10 to 20 degrees ipsilateral to direction of head rotation (e.g. Berthotz, 1981; Roucoux et al, 1981). The combined head and eye movement recenters a target in the visual field.

Animals who possess a rather homogeneous retina visually scan their environment with combined head movements and "reorienting" saccades (Van der Steen & Collewijn, 1984). These quick eye movements presumably evolved as a mechanism for afoveate creatures to reorient to targets on the edge or outside of their visual field with synchronous eye and head movements (Robinson & Zee, 1981). These occur only in conjunction with head reorientation (Collewijn, 1977; Easter et al, 1974) in the direction of interest. Like nystagmus quick phases, these saccades are craniotopically organized. The eye rotates toward a particular position in the orbit 5-10 degrees ipsilateral to the direction of head movement. Both quick phase and orienting saccades tend to direct the eye to highly predictable positions, which seem to be determined by head motion.

Animals have evolved whose survival depended on an ability to locate and secure a mobile source of food. Predators required the ability to identify and localize prey. As a result, the retina of visually dominant predators often has a region specialized for the resolution of detail and spatial discrimination, an area centralis or fovea. As the retina became more specialized, oculomotor control had to follow suit to compensate for the decreased field of this specialized region of the retina. Retaining their ability to shift the eye quickly and with a minimum of visual disruption to a particular position, saccades became more precise and their final craniotopic position highly variable. Something of potential interest would be selected in the peripheral field and gaze rapidly shifted to position that interesting image into central retina for further scrutiny. The direction and extent of eye motion would be determined by the distance of the target image from central retina. Thus, the reference position or goal of these saccades could be any craniotopic position within the oculomotor range. Moreover, in order to escape detection by the prey, head motion needed to be minimized during these shifts in gaze.

This latter, third category of quick eye movements, the foveal saccade, has been postulated as distinctive from quick phase and orientation saccades in two ways; they do not require head motion to complete a shift in gaze, and they are retinotopically organized (Robinson & Zee, 1981). The implication of numerous stimulation studies has been that foveate saccades are driven by a retinal error signal that is relative to the fovea (e.g. Roucoux, Crommelinck & Guitton, 1981; Schiller & Koerner, 1971). For example, stimulation of a region 10 degrees up and to the right of the fovea, stimulates a corresponding region in the superior colliculus to initate pre-motor activity in the deep motor layers of that same structure. The final result is a saccade 10 deg up and to the right to position the image of the target within or very near the fovea. Such a sequence progresses regardless of eye position and thus appears organized almost completely within an oculocentric or retinotopic framework. Puzzling is the approximately 120-130 ms latency between the discharge of sensory neurons in the superficial layers and the discharge of motor neurons in the deeper layers (Sparks & Mays, 1981), a latency that implies extensive multisynaptic processing.

FIXATION EYE MOVEMENTS.

The OKN and VOR systems keep the eye reasonably stable in the head, and a moving visual field reasonably well stabilized on the retina. In addition to these mechanisms, the foveate creature needs to maintain the image of a moving object within central retina. Foveate tracking systems evolved which involved not only small, foveate saccades, but also smooth, slower velocity eye movements which mimicked the target motion and thus kept it sufficiently stable on the retina for accurate identification and localization. Moreover, smooth eye movements work in conjunction with saccades to correct deviations of a target image from the fovea (e.g. Nachmias, 1959). These systems again likely co-evolved with the differentiation of central retina for the resolution of detail.

Little is known about the fixation eye movements of afoveate creatures. The rabbit, for example, has a horizontal visual streak which it likely maintains in alignment with a horizon if one is visible. Such alignment involves head motion (Van der Steen & Collewijn, 1984). The reorienting saccade approximately centers the eye within the orbit, reducing the amount of head motion necessary to recenter a scene of interest within the visual field. On the basis of our understanding of orienting saccades, this fixation mechanism is likely craniotopic rather than retinocentric.

IMPLICATIONS FOR PEOPLE WITH CENTRAL RETINAL DISEASE.

Robinson & Zee (1981) have proposed that we retain three types of quick eye movements; quick phase, reorienting saccades and foveate saccades. Implicit in many discussions of foveal saccades and eye movements is their essential retinotopic nature. The various mechanisms for control are initiated by a signal that measures the distance of a target image from a reference point assumed to be the fovea. The resultant eye movement reduces or eliminates that error by directing gaze at the target, shifting the image to the fovea. This assumption has rather devastating implications for many of those afflicted with a leading cause of permanent vision loss in the U.S., macular disease.

Although macular disease often destroys our region of highest visual acuity, most of the retina is left intact. The practitioner can compensate for the loss in visual acuity through the use of magnification systems. A person disabled by maculopathy must now direct target images to a functional region of peripheral retina. If indeed the "foveate" saccade and fixation systems were retinotopic and foveally referenced, these individuals would confront an intractable tendency to foveate objects of interest both following saccades and during fixation. If this were the case, an optimum rehabilitation strategy would be to suppress the foveal saccade and fixation systems and depend on orienting saccades to keep the eye in a particular, craniotopically referenced position. The optimum eccentric gaze angle would be more precisely controlled with head movements in the manner of afoveate animals. We have found in our own low vision services that people with central vision loss often exhibit a tendency to twist their head to view a target and often use considerable head motion during reading. At least one rehabilitation approach is implicitly based on this model. Prism scanning, involves deviating a target image to the peripheral retina with prism while the patient maintains an approximately straightforward viewing posture (Romayanda et al, 1982). Under this model such activities that involve frequent, rapid shifts in gaze such as reading would be prohibited by the sheer inertia of the head. The prognosis for rehabilitating visual functions that require considerable scanning such as reading would be poor.

FOVEAL SACCADES WITHOUT VISION

An alternative model would be that scanning and fixation without head motion, that is the so called "foveate" saccade system, is not necessarily controlled by a foveally referenced retinal error signal. Either the retinal reference point is shifted, or the system is most fundamentally controlled within an egocentric rather than oculocentric framework, much like our other motor control systems. For example, we are able to localize objects with hand motions on the basis of non-visual cues, motor control that must combine sensory information with information about body position.

Several lines of research suggest that all saccades (foveal, orienting and quick phase) are controlled essentially within a craniotopic, that is, egocentric framework. As an example often used in the rarefied atmosphere of the laboratory, a monocular headcentric framework is set up by patching one eye and afixing the head. People are able to saccade to and maintain various eye positions in complete darkness (Skavenski & Steinman, 1970; Becker & Klein, 1973) or follow the predicted trajectory of a target without retinal input (Whittaker & Eaholtz, 1982). These findings point to the existance of an oculomotor control mechanism that compares actual with learned or "desired" eye position, the desired eye position being a reference position. In complete darkness an arbitrary reference position is learned; they saccade to the remembered position of a visible target. Several neurophysiological studies confirm that at the collicular level, saccade trajectories can be programmed and controlled without a retinal error signal (Becker et al, 1981; Sparks & Mays, 1983; Sparks & Porter, 1983). In these studies, monkeys were trained to saccade to lights presented peripherally. During the saccade or near its onset, collicular stimulation displaced the eye in a different direction than intended. The saccade, however, terminated with the eye in its intended position, sometimes with mid-course correction in the perturbed trajectory. Thus a feedback system must be continually operating during a saccade. Since retinal feedback can be ruled out as being excessively delayed, we are left with a system that compares actual eye position with desired eye position, to compare eye possition error (Robison, 1873) . Sparks & Porter (1983) went on to document that intermediate and deeper layers of the superior colliculus encode motor error (error within a monocular egocentric framework) rather than necessarily retinal error.

With normally sighted foveate primates under normal monocular viewing conditions motor error would closely correspond to the retinotopic map apparent in the overlying superficial layers of the superior colliculus, the reference or "desired" eye position being defined by the visual axis. However, in the absence of retinal input or dependence on other sensory information, an eye position error signal can be generated. The motor control and presumably the foveal saccade machinery can operate in the absence of direct retinal control. If, however, a target is continuously visible, a resultant retinotopic signal would activate brain stem localization mechanisms. In this case will a foveally referenced retinotopic signal dominate an attempt to saccade to a visible target even in the absence of a functioning fovea?

"FOVEAL" SACCADES AND FIXATION WITHOUT A FOVEA

Foveal and orienting saccades can be distinguished by the involvement of head motion. Only foveal saccades can occur in the absence of head motion. We questioned whether visually guided eye movements would persist in directing the image of a target toward central retina even when it was scotomatous.

Individuals with a variety of macular pathologies were selected for study including Stargardt's, atrophic, and exudative macular degeneration. Two cases were monocular pathologies. Following informed consent, and evaluations of ocular health, fixation eye movements and saccades were objectively evaluated.

Eye movements were measured using a search coil in an electromagnetic field. The fields were generated by 6 foot coils and the search coil tightly affixed to the eye with the Skalar silicon annulus (Collewijn et al, 1975). The head was restrained, usually with a bite bar, but occasionally with a forhead and chin rest. Following objective calibration of a search coil, it was positioned on the eye. The system sensitivity was better than 3 minarc over a 38 degree positional range. The bandpass was limited by the 200 Hz sampling rate to 0-100 Hz. All data were digitally stored.

After one eye was patched and refractive error corrected subjects were asked to fixate single characters that were sized in accordance with their best visual acuity. In some cases they were asked to read a line of characters repeatedly, during other trials a subject was asked to first look toward the extreme corner of the rear projection screen and look at the target immediately following its onset so that it was clearest . At the end of the session, central fields were determined with projection perimetry. Central absolute scotomata were defined with a 65 ml, 4 minarc target on a less than .007 ml background. Highly reliable fields were obtained despite unstable fixation through the use of an image stabilizing system that compensated for eye movements. High speed mirror galvanometers were yoked to the horizontal and vertical eye position signal to stabilize the target on the retina. The test target could then be systematically moved over the surface of the fundus.

Most of our subjects with macular pathologies whom we have examined thus far can saccade to and maintain an eccentric eye position that positions the image of a target of interest out of the central scotoma. Figure 1 illustrates 2 subjects who represent extremes in performance. Subject FC had a monocular central scotoma, and was unable to maintain a target image out of the central scotoma when her better eye was patched. Subject DM had a bilateral central scotoma resulting from Stargardt's macular degeneration and typified subjects with scotoma less than 20 degrees in her ability to maintain the target image within a preferred eccentric viewing locus out of the central scotoma.

Eccentric fixation actually can be readily learned. We created an artifical scotoma by monitoring the retinal image position of a normally sighted individual. When the target image fell within a predetermined distance from the fovea, a shutter obscured it. Nothing else was visible to the subject, leaving him to depend entirely on the target image to direct his eye movements. Within a few 17 second trials this subject, who was engaged in a visual acuity task, was able to reliably maintain the target image out of the scotoma with small corrective saccades (figure 2). This same subject continued the visual acuity task with the same sized scotoma for 53 trials over a number of daily sessions. At the end, he did not exhibit a preference for a particular gaze angle or retinal image position; rather the target image was usually directed to a number of positions around the edge of the scotoma, again with small saccades and slow eye movements. Three normally sighted subjects demonstrated a capacity to contain (68% of) these searching eye movements within an approximately 2400 to 3100 minarc area simply by asking them to hold their eye as still as possible. Thus, even with a visible target, subjects with intact fovea are able to readily redefine the reference ("desired") eye position for the control of corrective saccades and the maintenance of fixation.

One distinction between foveal and extrafoveal fixation, is in the distribution of the fixation eye movements. For the normally sighted, fixation can be closely approximated as a unimodal bivariate distribution (Nachmias, 1959) indicating a single reference position. Eccentric fixation patterns are multimodal both

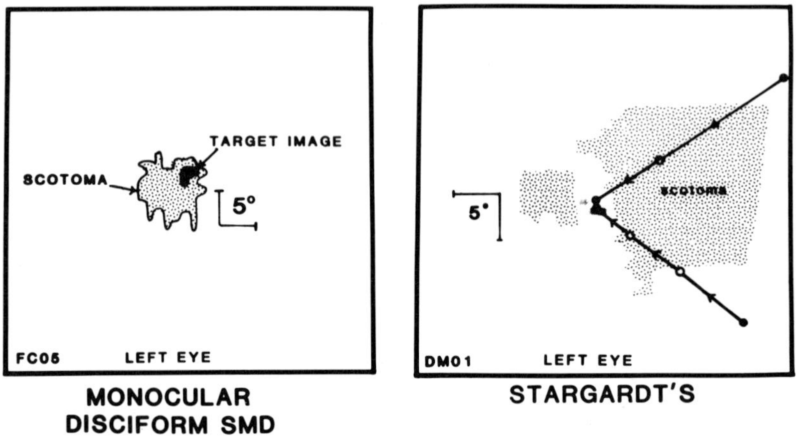

Figure 1. Field projection of the central scotomata (stippled area) and the region where the image dwelled for the highest 67 percentile of the 12 second fixation period (cross-hatched and filled areas combined). The region where the image dwelled for the highest relative time (accumulating to 34%) is indicated by the filled area. Subject FC (left) had a monocular scotoma. Subject DM had binocular scotomata. The smaller scotoma in the left panel represents the physiological blind spot. Straight lines connect the onset and termination of saccades that occured at the trial start with the open circles marking fixation periods shorter than 100 ms.

in normally sighted with an artifical scotoma and subjects with a macular scotoma as illustrated in figure 3, suggesting unsystematic shifts in this reference position. We have found that fixation variability usually increases when subjects change from spot fixation to character recognition and we suspect fixation variability represents, in part, scanning type eye movements that are compensatory for the low spatial resolution of the retina.

Subjects with central scotoma and restrained heads were able to direct targets over some distance to the preferred, eccentric viewing locus. For example DM (figure 1) was able to saccade directly to her preferred eccentric fixation postion. Somewhat unusual about her saccades was a tendency to string hypometric saccades together. The open circles in this figure indicate fixations of less than 100 milliseconds. Typically, a minimum fixation duration of at least 116 ms is necessary to aquire, process and respond to visual information with saccades (for review see Vaughan, 1983). The saccade following the shorter fixation period must have been programmed prior to the fixation and is more likely, in part, a predetermined strategy.

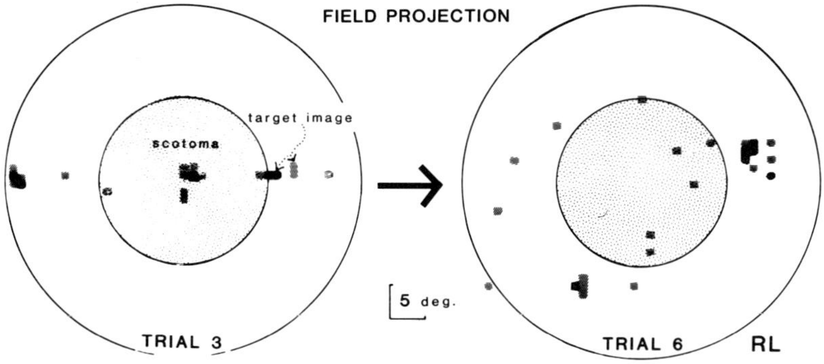

Figure 2. Field projection of an artificial central scotoma (stippled area) and fixation dwell areas (see fig. 1 caption) of a normally sighted young adult. Performance during the last 12 seconds of the third and sixth 17 second trial are represented.

Hypometria has been associated with reduced or distorted visual information such as saccades to auditory targets without visual information (Traccis et al, 1984), homonymous hemianopsia (Meienberg et al, 1981) or saccades generated prior to foveal maturation (e.g. Roucoux et al, 1983). In addition, a similar eye movement pattern has been associated with neurological disease affecting cerebellar function (e.g. Baloh et al, 1978). With scotoma this strategy could (in part) represent a conservative strategy of extreme undershooting to avoid losing a target image in scotoma (Meienberg et al, 1981); however, it may also represent a response to a breakdown in encoding retinal error and desired saccade trajectory. If direction and error information were encoded at no better than an ordinal level, or with significant random or systematic error, then a strategy of repeatedly sampling would be necessary in order to attain the intended eye position.

Subject FC was reexamined approximately 4 months after central vision in the right eye was lost, creating a bilateral loss in central vision, and was in a much earlier stage of adaptation than our other subjects. She still generated apparently foveally directed saccades (figures 4 & 5 right eye) which in her case left the target image in absolute scotoma. The saccade patterns seen in her other, preferred, eye (figure 4 & 5 left eye) exhibited a

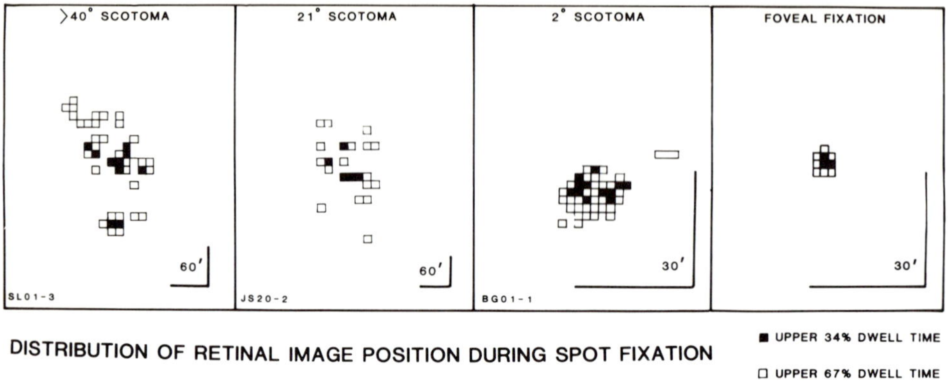

Figure 3. Detailed representation of the distribution of a target image on the retina of three subjects with binocular central scotomata and foveal fixation of a normally sighted young adult. Different scotoma sizes (mean diameter indicated on top) resulted from exudative macular degeneration (SL), Stargart's macular degeneration (JS), and a macular hole (BG). Subjects were instructed to fixate a spot "so that it was seen most clearly".

pattern more typical of patients with bilateral macuolopathies whom we have evaluated and likely represented an intermediate stage of adaptation. The difference between eyes results from a non-correspondence in the preferred viewing areas between eyes, preventing FC from using both eyes at the same time. She was typical of patients with bilateral central scotoma in that she reported a strong preference for using one eye, the eye which had adapted more to her vision loss.

JS, a 24 year old woman who had been afflicted with a bilateral maculopathy for over 5 years, apparently had considerably more time to adapt to her vision loss. Large saccades were common and were able to shift the target image directly to her perferred viewing area. Although she did exhibit hypometria, the intersaccade fixations were less then 100 ms (see also her sister, DM in fig. 1). The smaller saccades, typical of visual scanning and reading, likewise tend to direct the image of interesting targets to a locus out of the foveal region. When asked to reread a string of characters to herself, she exhibited the staircase scanning pattern typical of a normal, foveate individual (figure 6). Instead of directing the target to the fovea, however, she directed it to an extrafoveal reference position.

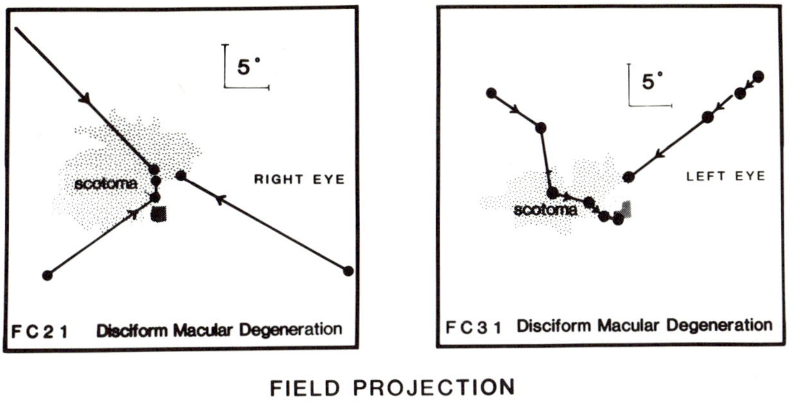

Figure 4. Field projections of the macular scotomata (stippled area), fixation dwell areas (see fig. 1 caption) and saccades (solid lines) that occured at the start of the trial. Solid circles represent fixation positions that separated the saccades. Data are from each eye of a subject with exudative macular degeneration. The right eye lesion including the effects of Argon laser photocoagulation efforts was approximately 4 months old.

Our observations of fixation eye movements and saccades of people with macular scotoma indicate that these eye movements can direct a continuously visible target image to a non-foveal locus and maintain the image in this locus. These eye movements did not involve head motion, and were apparently partly in response to retinal information, and thus resembled "foveate" fixation and saccadic eye movements. Larger saccades might have been directed partly by learned eye position since the targets were projected within an 10 X 10 deg. area on the screen. Smaller refixation and scanning saccades, however, are likely a response to retinal error signals. In any case, no overriding, reflexive tendency to foveally fixate was apparent. These fixation and saccadic eye movements can, therefore, function to maintain a target image in an arbitrary retinal locus.

Although subjects with macular scotoma have demonstrated the capacity to shift the retinal locus of control to an extra-foveal area, the generation of saccades is apparently not readily aquired as an unconscous habit. Indeed certain parallels exist between the initial, gradual development of foveate saccades and the redevelopment of saccade control following loss of foveal function.

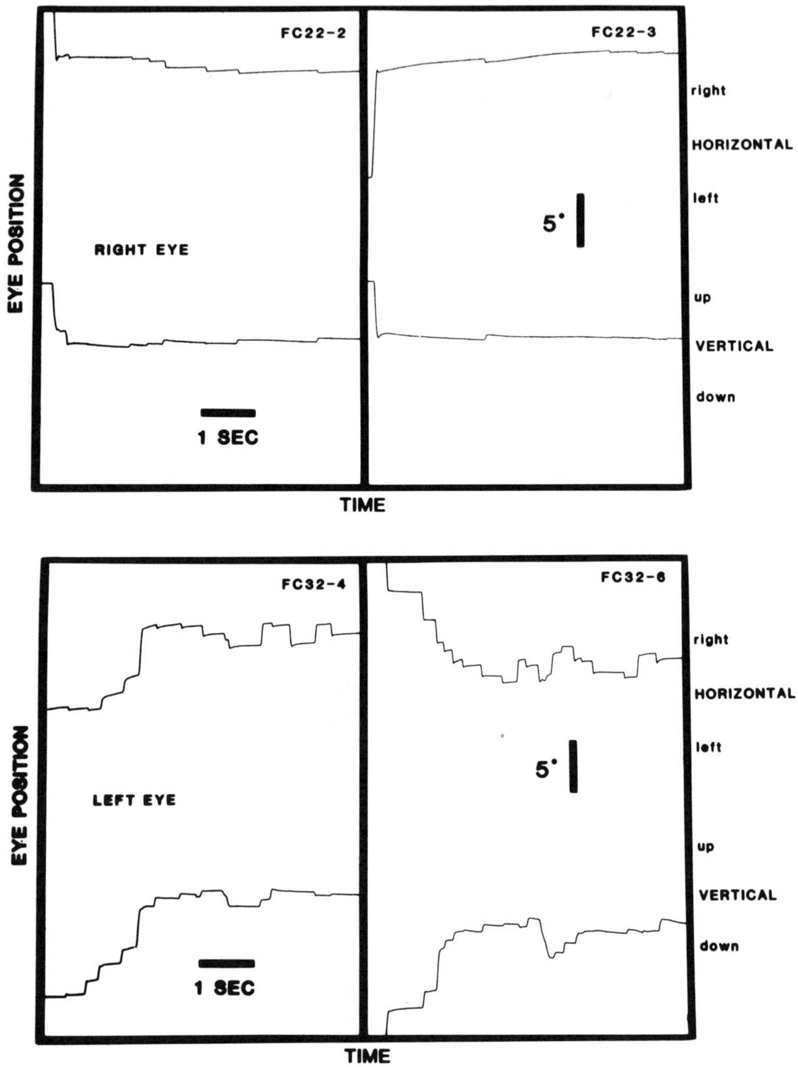

Figure 5. Eye position is plotted as a function of time. The subject was attempting to shift gaze by more than 10 degrees in response to an instrunction to "look at the target so that it is seen most clearly". These same data are also represented as two of the saccades in each panel of figure 4.

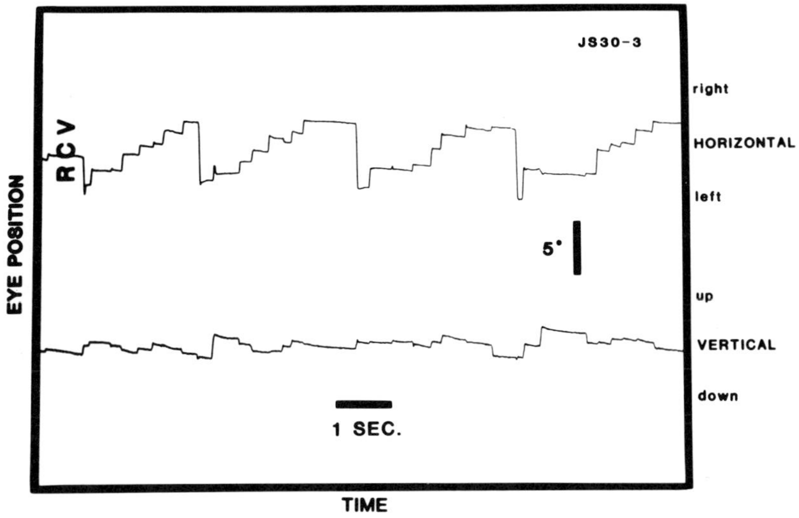

Figure 6. Eye position is plotted as a function of time. The subject is afflicted with bilateral Stargardt's macular degeneration and a central scotoma averaging approximately 20 degrees in diameter. She is following instructions to read the line of characters over and over again to herself.

Initially the neonate seems to regularly use head movements to shift gaze; that is "orienting saccades" are more common, later giving way to more frequent "foveate" saccades (Roucoux et al, 1983). Apparent improvements in saccade programming accuracy develop as shifts in gaze are completed with one or two saccades rather than strings of hypometric saccades (Roucoux et al, 1983). The often legally blind, adult who loses central vision may very well be forced to embark on that long, undoubtably tedious, journey which we have all coursed. Our findings suggest that those who are not frustrated may hope to reacquire an ability to rapidly scan visual material and thus enjoy a substantial recovery of visual function.

ACKNOWLEDGMENTS

This research was supported by NEI grant # EY04754, NIHR grant #G008435033 and grants from the Pennsylvania Lions Sight Conservation and Eye Research Foundation. We would like to thank our colleagues at the William Feinbloom Vision Rehabilitation Center for their many insights. Robert J. Ellis, Sr. designed and fabricated many critical components of the instrumentation described herein. Janis White Initially suggested that with central scotoma, the locus of controf of foveal saccades might shift.

REFERENCES

Baloh, R.W., V. Honrubia and A. Sills. 1977. Eye-tracking and optokinetic nystagmus. Ann Otol, 86, 108-114.

Becker, W. and H.M. Klein. 1973. Accuracy of saccadic eye movements and maintenance of eccentric eye position in the dark. Vision Res., 13, 1021-1034.

Becker, W., W.M. King A. Fuchs R. Jurgens G. Johanson and H. Kornhuber. 1981. Accuracy of goal-directed saccades and mechanisms of error correction. In: Progress in Oculomotor Research. A. F. Fuchs and W. Becker (Eds.), Elsevier, North-Holland, 29- 38.

Berthoz, A., 1981. Discussion summary (Metrics and Models of Saccadic Eye Movements). In: Progress in Oculomotor Research. A. F. Fuchs and W. Becker (Eds.), Elsevier, North-Holland, 53-56.

Collewijn, H., F. van der Mark and T.C. Jansen. 1975. Precise recording of human eye movements. Vision Research, 15, 447-450.

Collewijn, H., 1977. Eye and head movements in the freely moving rabbit. J. of Physiol. (Lond.), 266, 471-498.

Dobson, V., 1983. Preferential looking measures of visual acuity in infants and young children. In: Molecular and Cellular Basis of Visual Acuity. S. R. Hilfer and J.B. Sheffield (Eds). Springer-Verlag, New York, 151-172.

Easter, S.S., P.R. Jr. Johns and D.R. Hechenlively. 1974. Horizontal compensatory eye movements in the goldfish (Crarassius auratus): I. The normal animal. J. Comp. Physiol., 92, 23-35.

Meienberg, O., W. Zangemeister M. Rosenberg W.F. Hoyte and L. Stark. 1981. Saccadic eye movement strategies in patients with homonymous hemianopia. Ann. of Neur., 9, 538-544.

Nachmias, J., 1959. Two-dimensional motion of the retinal image during monocular fixation. J. Opt. Soc. Am., 49, 901-908.

Robinson, D.A., and D.s. Zee. 1981. Theoretical considerations of the function and circuitry of various rapid eye movements. In: Progress in Oculomotor Research. A. F. Fuchs and W. Becker (Eds.), Elsevier, North-Holland, 3-10.

Robinson, D.A., 1973. Models of saccadic eye movement control system. Kybernetik 14, 71-83.

Romayanda, M., S.W. Wong I.H. Elzeneiny and G.H. Chan. 1982. Prismatic scanning method for improving visual acuity in patients with low vision. Ophthalmol.,89, 937-945.

Roucoux, A., M. Crommelinck and D. Guitton. 1981. The role of superior colliculus in the generation of gaze shift. In: Progress in Oculomotor Research. A. F. Fuchs and W. Becker (Eds.), Elsevier, North-Holland, 129-136.

Roucoux, A., M. Crommelinck J.M. Guerit and M. Meulders. 1981. Two modes of eye head coordination and the role of the vestibulo-ocular reflex in these two strategies. In: Progress in Oculomotor Research. A. F. Fuchs and W. Becker (Eds.), Elsevier, North Holland, 309-318.

Roucoux, A., C. Culee and M. Roucoux. 1983. Development of fixation and pursuit eye movements in human infants. Behav. Brain. Res. 10, 133-139.

Schiller, P.H., and F. Koerner. 1971. Discharge characteristics of single units in the superior colliculus of the alert Rhesus monkey. J. neurophysiol. 34, 920-936.

Skavenski, A.A., and R.M. Steinman. 1968. Control of eye position in the dark. Vision Res., 10, 193-203.

Sparks, D.L., and L.E. Mays. 1981. The role of monkey superior colliculus in the control of saccadic eye movements: A current perspective. In: Progress in Oculomotor Research. A.F. Fuchs and W. Becker (Eds.), Elsevier, North-Holland, 137-144.

Sparks, D.L., and L.E. Mays. 1983. Spatial localization of saccade targets. I. Compensation for stimulus induced purturbations in eye position. J. Neurophysiol. 49, 45-63.

Sparks, D.L., and J.D. Porter. 1983. Spatial localization of saccade targets. II. Activity of superior colliculus neurons preceding compensatory saccades. J. Neurophysiol. 49, 64-74

Traccis, S., L.A. Abel and L.F. Dell'Osso. 1984. Audio-ocular response: saccadic programming. Aviation Space Env Med, 55, 734-739.

Van der Steen,J., and H. Collewijn. 1984. Ocular stability in the horizontal, frontal and sagittal planes in the rabbit. Exp. Brain Res. 56, 263-274.

Vaughan, J., 1983. Saccadic reaction time in visual search. In: Eye Movements in Reading: Perceptual and Language Processes. K. Rayner (Ed). Academic Press, New York, 397-409.

Whittaker, S.G., and G. Eaholtz. 1982. Learning patterns of eye motion for foveal pursuit. Vision Research. 23. 393-397.

Abnormal Early Visual Experience and Visual Function in Monkeys

M.L.J. Crawford

The relationship between anomalous early visual experience and visual dysfunction has been noted in clinical observations for the past century, and yet it has been only in the past 25 years that experimental manipulation has begun to provide the solid physiological and behavioral data needed for an understanding of the process of visual development. Outstanding amongst the essential animal models for human visual processes has been the Macaque monkey, being demonstrated to be quite comparable to man in visual physiology, anatomy, spectral sensitivity, contrast sensitivity, and binocular functions. Moreover, the sensitivities of young monkeys can be altered by early abnormal visual experience in ways to mimic the consequences often seen in the pediatric eye clinic, thereby qualifying this monkey as the best model for research on the underlying physiological changes associated with abnormal visual experience in early life.

As this symposium is concerned with the development of order in the visual system, it should be noted that at birth the connections of the visual system of the monkey are already in order; are developed and ready for the instructions of experience to define and shape the functions of the nervous system to that of the finely-tuned, highly-sensitive visual system of the adult primate. When the two eyes are normal and ordinary visual experience is permitted, the complex process of visual development proceeds with deceptive simplicity. In the visual brain, the enormously large number of synaptic contacts there undergoe tremendous reorganization, or reordering, at the instigation of visual experience. Congenital order in the visual system is thereby changed to reflect the functional order developed by the sort of visual experience attendant during the first weeks and months of life.

There are, however, many points of vulnerability in the course of visual development. As is now well known, if there is a congenital imbalance between the two eyes such that one eye is placed at a disadvantage, that eye loses control over neurons of the visual system, and may become functionally blind. Alternatively, if the two eyes are given abnormal experience such as misalignment of the visual axes, the binocular neurons disappear from the visual cortex, binocular functions are lost and do not recover with time. Finally, the normal spectral sensitivity functions of the primate eye are permanently altered if monocular occlusion is continues past the first few weeks of life.

It is the intention here to review and summarize some of the work we have done concerning the effects of abnormal early visual experience on the visual capacities of Macaque monkeys. (The "we" referred to hereafter in this paper are: Drs. Ron Harwerth and Earl Smith, of the University of Houston and Dr. Gunter von Noorden of the Baylor College of Medicine in Houston.) We are convinced that the visual processes and their underlying physiological mechanisms seen in the monkey have direct qualitative parallels in the developmental events and processes of vision in children.

During the past few years, we have simulated many of the conditions often encountered in pediatric vision clinics in young rhesus monkeys and this presentation summarizes conclusions from several of our experiments. The behavioral tests have assessed 1) the spatial modulation sensitivity (Harwerth, et al., 1981, 1982a,b), 2) temporal modulation sensitivity (Harwerth, et al., 1982), 3) spectral sensitivity (Harwerth, et al., 1981, 1984), and 4) the binocular sensitivity or stereopsis (Crawford, et al., 1983) of monkeys reared with 1) monocular form deprivation, 2) surgically or optically induced strabismus, and 3) pharmacologically or optically induced anisometropia. Our physiological approach has been to record from individual cells of the visual cortex, describing the stimulus requirements for evoking the best response, and then classifying the neurons according to the well-known 7-category eye-dominance scheme (Hubel & Wiesel, 1962).

The methodological approach has been to alter the visual experience of infant monkeys within the first few weeks of life (von Noorden & Dowling, 1970; von Noorden, Dowling & Ferguson, 1970; Harwerth, et al., 1981; Crawford, et al., 1983) and subsequently test their visual sensitivities using behavioral procedures. Subsequently, we have studied the electrophysiological and anatomical conditions of the visual brains of these same animals (Crawford, et al., 1975 and 1984); only some results of the physiological studies being described here.

Beginning at about 1 year of age, normal and experimental monkeys were trained in a variety of visual experiments. Seated in a primate chair, the monkeys were trained to press and hold down a lever at the start of each trial which was signaled by the onset of an auditory cue. The task of the monkey was to watch a visual display and to release the lever whenever there was a change in the display. Three kinds of visual display have been used, although the behavioral task has been essentially the same with each display. An oscilloscope display (Campbell & Green, 1965) has been used whenever spatial frequency modulation or temporal modulation sensitivities are measured. When color sensitivity is measured (Harwerth, et al., 1981), the display has been a circular white visual field presented in Maxwellian view, upon which a spectral test flash is superimposed. And finally where stereopsis has been tested (Crawford, et al., 1983), the visual display has been the face of a color television screen upon which anaglyphic red and green dynamic random dot stereograms are presented. In short, these tests for spatial vision, color vision, and binocular vision have all been threshold detection experiments.

Our first studies were to address the question as to what happens to spatial visual sensitivity following a variety of abnormal early visual experiences. Examples of some of the results of these experiments are summarized in the following figures. When the contrast sensitivity for sine wave grating patterns is measured in a group of normal human or normal monkey observers (Harwerth, 1982) a characteristic curve is obtained as shown in figure 1. First the human and monkey peak contrast sensitivity for sine wave gratings is about 3-4 cycles per degree visual angle, with sensitivity falling off toward higher as well as toward lower spatial frequencies. The average contrast sensitivity high spatial frequency cutoff or approximate maximum resolution for human observers is approximately 44 cycles per degree as compared to approximately 32 cycles per degree for normal monkey observers. The shapes of the normal contrast sensitivity functions are approximately the same for both, with the human curve exhibiting

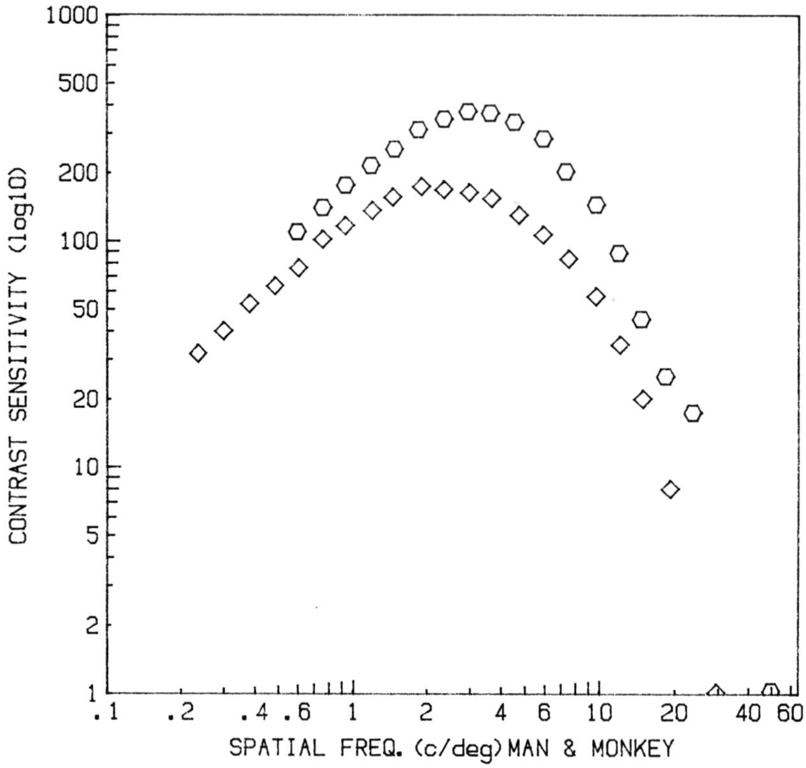

Figure 1. Average contrast sensitivity functions for 12 human observers (hexigons) and the normal eye of 12 rhesus monkeys (diamonds). Extrapolated high spatial frequency cut-off values are shown on the abscissa (see Harwerth, 1982).

about a half log unit higher sensitivity. Motivational levels probably account for these slight differences. It is clear from examining these two curves that the functions would be compatible with common underlying mechanisms.

The consequences of monocular form deprivation are illustrated in figure 2. The results of long-term eyelid suture are seen as a profound depression of contrast sensitivity; the results of which would be equivalent to finger counting or blindness in that eye. In this particular case two monkeys had the eyelids of one eye closed for 18 and 22 months beginning at approximately 30 days of age. The contrast sensitivity functions for the non-deprived eyes match very well that obtained in normal monkey observers whereas those for the

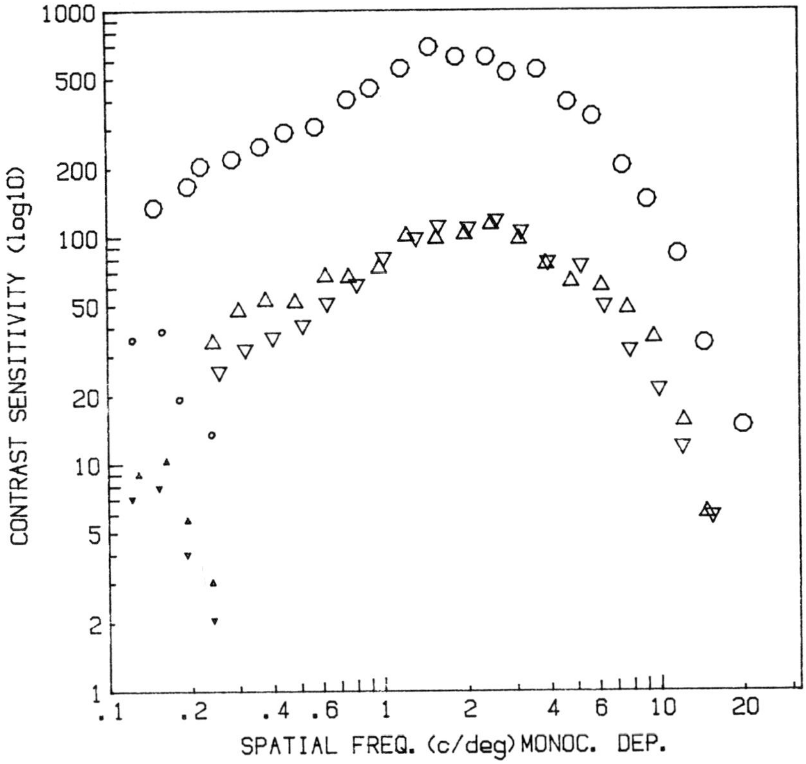

Figure 2. The effects of monocular deprivation by eyelid suture of long duration (18-22 mo.) is contrasted between the normal eyes (large triangles) and deprived eyes (small triangles) of two monkeys. Profound amblyopia is indicated by the 0.3 c/deg cutoff. Short-term eyelid suture of only 2-weeks can lead to comparable losses as shown for a third monkey (large circles, normal eye; small circles, deprived eye).

deprived eyes are dramatically reduced. Shown on this same figure are the consequences of short-term monocular form deprivation for another monkey where the eyelids were closed for only two weeks. Notice that this result is quite comparable to that for the long-term monocular closure. In short, monocular eyelid closure for as brief a period as two weeks beginning early in life can produce an irreversible deep amblyopia in the occluded eye.

In a manner similar to that shown for monocular lid closure and consequent monocular form deprivation, we carried out a series of experiments to evaluate the effects of eye misalignment by experimental surgery. Unilateral strabismus either in the form of a surgical esodeviation (esotropia), or a surgical exodeviation (exotropia) was produced in infant monkeys and examples of the results are shown in figure 3. Esodeviation produces a level of severe amblyopia comparable to that shown above for monocular lid suture, in that the deviated eye exhibits a very poor acuity comparable to that obtained with eyelid suture. It should be pointed out that the degree of eye misalignment in this particular monkey was quite severe and we have yet to quantify the degree of acuity loss with a modest degree of eye misalignment. Exodeviation, as also shown by the example in figure 3, produces a clear but less severe consequence for contrast sensitivity when compared to the effects of esotropia as just described. In summary, spatial vision sensitivity is dramatically reduced by either eyelid closure (monocular deprivation) or eye deviation by either an esotropia or an exotropia.

In addition to the consequences for spatial vision, these particular forms of abnormal early visual experience have adverse effects upon the temporal modulation sensitivity of the visual system (Harwerth, et. al., 1983). Monkeys were trained to watch an oscilloscope screen and detect when the screen began to flicker. The luminance of the screen was sinusoidally modulated about the mean luminance (40 cd/m^2) to produce a uniform field flicker. Figure 4 shows examples of the contrast modulation sensitivity as a function of temporal frequency for two monocularly lid-sutured monkeys. Monocular deprivation produces a reduction in the contrast modulation threshold at all temporal frequencies and reduces the high temporal frequency cutoff from about 64 Hz down to approximately 32 Hz. This severe loss occurs as well in an eye subjected to monocular deprivation for a period of only two weeks (Harwerth, et al., 1983b, fig. 3). Taken together, it is clear from these examples that either abnormal eye alignment or abnormal monocular occlusion in young monkeys produces a dramatic reduction in both spatial and temporal visual sensitivity. Moreover, it should be stressed that these adverse effects become clear within a matter of a few weeks of abnormal experience.

In an effort to simulate another type of amblyopia, we have manipulated the quality of the image in the early visual period of young monkeys (Crawford, 1977; Harwerth, et al., 1983). One technique was to chronically apply atropine to one eye of a young

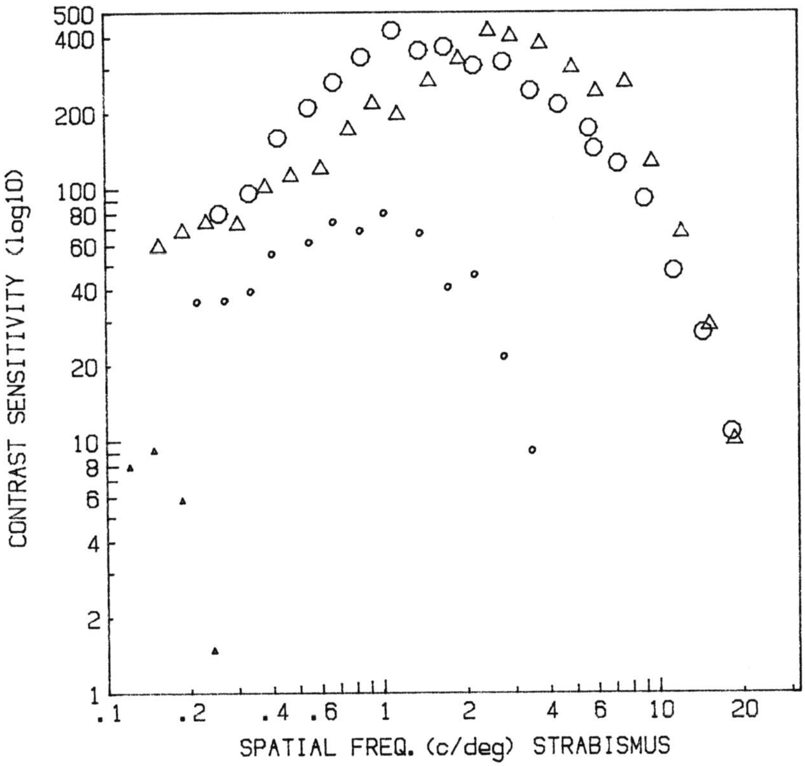

Figue 3. The effects of surgical strabismus upon contrast sensitivity of the normal eye (large symbols) and deviated eye (small symbols) for esotropia (triangles) and exotropia (circles). Note that the esotropic eye is deeply amblyopic (small triangles) while in the second monkey with exotropia (small circles), the amblyopia is clear, but less severe.

monkey throughout the first few weeks and months of life. Atropine sulfate was dropped into one eye each day so that a chronic state of cycloplegia was maintained. Moreover, to assure that the image of the atropinized eye was blurred at all times, the animal was kept in a white lighted box of about 1 m^3 so that nothing could be viewed beyond 1 meter viewing distance. As shown in figure 5 chronic cycloplegia produces an attenuation of contrast sensitivity in the treated eye resulting in a milder form of amblyopia as compared to the previous results obtained by monocular deprivation or by strabismus. Our most recent efforts to create anisometropia,

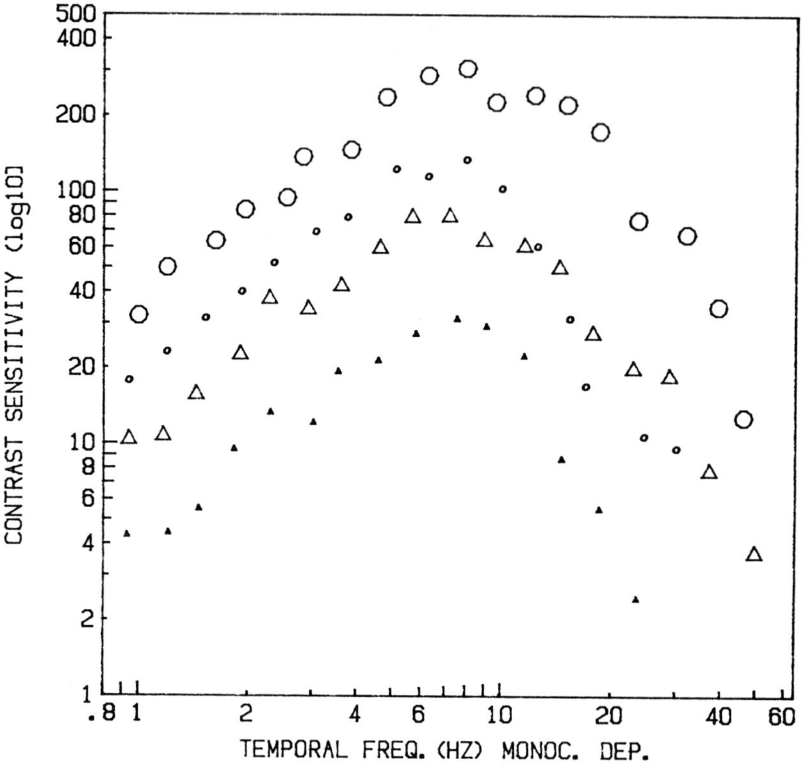

Figure 4. Temporal contrast sensitivity functions for the normal eyes (large symbols) and eyes deprived by monocular lid-suture (small symbols) for 22 mo. (triangles) and 2-weeks (circles). Both experimental eyes show a significant difference in threshold flicker sensitivity relative to their normal eye.

involve using optical blurring techniques. Young monkeys were fitted with a helmet (Crawford & von Noorden, 1979; Smith, et al., 1985) for holding negative lenses in front of one eye for various periods early in life. Figure 5 illustrates the results of one such experiment along with the results of cycloplegia. The eye which viewed the world for 60 days through a -10 diopter lens is seen to be significantly reduced in contrast sensitivity with a high spatial frequency cutoff of only about 10 c/deg. The fellow eye by comparison had a normal high spatial frequency cutoff at approximately 22 c/deg. Experimental anisometropia by these two techniques results in sensitivity losses somewhat milder compared to

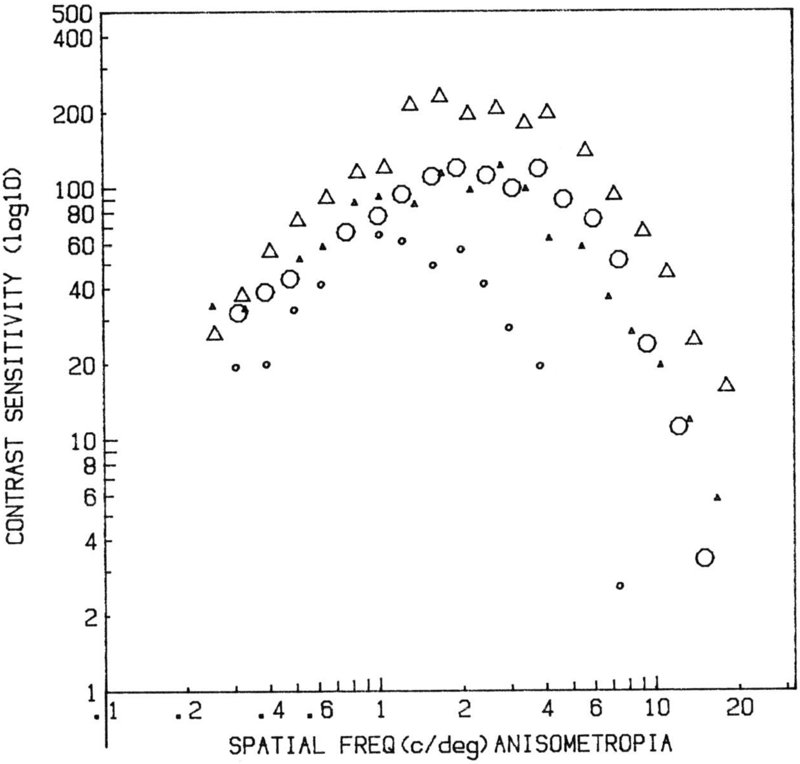

Figure 5. The effects of chronic monocular image blurring on spatial contrast sensitivity for the normal eyes (large symbols) and experimental eyes (small symbols) of two monkeys. Chronic cycloplegia (triangles) or optical anisometropia (circles) both produce reductions in spatial sensitivity.

those losses described above for monocular eye closure or strabismus. Current use of extended-wear soft contact lenses makes experimental optical anisometropia somewhat easier, both for the monkey and the researcher.

An unexpected consequence of early eyelid closure and monocular deprivation has been the effect upon spectral sensitivity (Harwerth, et al., 1981; 1984). For example, figure 9 shows several threshold spectral sensitivity curves under various adaptation conditions for the normal and lid-sutured eyes of a monkey. As previously described, the photopic spectral sensitivity of the normal monkey, like that of the human observer, is characterized by a broad spectral sensitivity curve having three peaks observed in the area of 610 nm, 535 nm and approximately 445 nm. These characteristic features of the photopic threshold spectral sensitivity curve (seen as the diamond symbols) have been described as the interaction between three different channels when measured upon a white background (Sperling & Harwerth, 1971 & Harwerth & Sperling, 1975). Now contrast in figure 6 the result of long-term lid suture upon the shape of the spectral sensitivity curve. Several effects of monocular deprivation are shown here. First, spectral sensitivity in the treated eye is on the order of three log units less sensitive than that of the companion eye under all conditions of adaptation. Second, and most important, the shape of the function from the deprived eye becomes unimodal and is adequately described by the shape of the rhodopsin curve corrected for preretinal absorption. Third, whereas the normal eye shows a 4-log unit shift in sensitivity with adaptation, the deprived eye shows only about 1 log unit shift with comparable changes in adaptation conditions. In short, long-term monocular eyelid closure not only produces a loss in spatial sensitivity or resolving power of the eye, a loss in temporal sensitivity functions, but also produces a substantial alteration in the spectral sensitivity function. The residual very poor threshold vision in the deprived eye is clearly determined by the rods. This result is somewhat surprising inasmuch as amblyopia is purported to be a neural dysfunction of the cortical brain, whereas color interactive mechanisms have traditionally been thought of as having loci within the retina itself.

In human subjects, it is not uncommon to find patients who have had congenital strabismus, subsequent realignment surgery and a regimen of therapy which is totally successful in either avoiding or correcting an amblyopia. However, such patients are often found to

Figure 6. Spectral sensitivity under four levels of light adaptation for the normal (4 upper curves) and monocularly deprived eyes (4 lower curves) of a monkey. All threshold curves are shown at relative sensitivity values on an absolute quantum scale. Adapting backgrounds were: Dark (circles and asterisk); 188 Troland (Td) background (triangles and X); 750 Td background (squares and +); 3,000 Td background (Diamonds). The solid lines represent the CIE scotopic luminosity in the top curve and the four bottom curves from the deprived eye. The solid curves through the three light-adopted curves for the normal eye were derived from a subtractive interaction model for color vision proposed by Sperling and Harwerth, 1971.

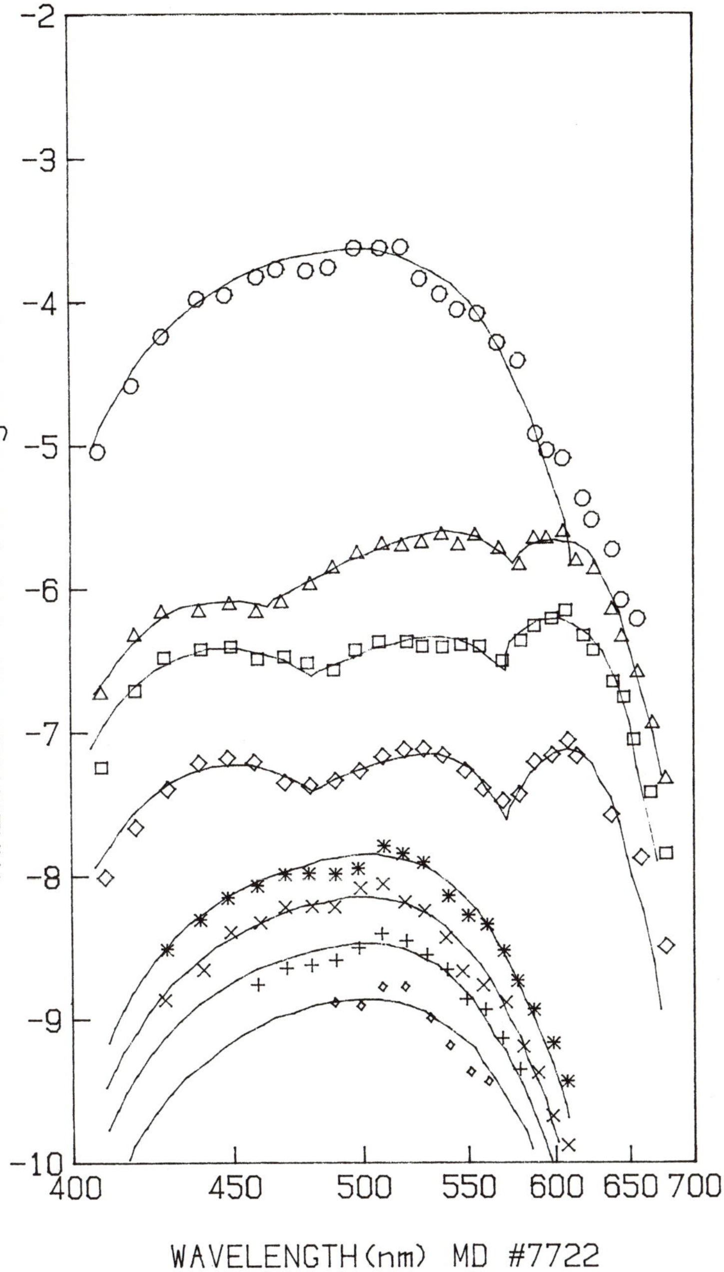
-2
-3
-4
-5
-6
-7
-8
-9
-10
THRESHOLD SENSITIVITY(log10 1/Q)
400
450
500
550
600
650
700
WAVELENGTH(nm) MD #7722

be weak in stereovision, or indeed found to be stereoblind. The physiological entity underlying stereovision or stereopsis, has been proposed to be the binocular cell of the visual cortex (Hubel & Wiesel 1962; Barlow, et al., 1967; & Bishop, 1973) In a series of studies we have demonstrated that the proportion of binocular cells in striate cortex can be reduced by subjecting the young monkey to optical dissociation of the visual images early in life (Crawford & von Noorden, 1979). Congenital binocular cells disappear from the striate cortex of young monkeys subjected to chronic dyplopia produced by viewing the world through optical prisms. The rate of disappearance of binocular cells is surprisingly rapid, decreasing to only about 10% of the normal population following optical dissociation of 60 days duration (Crawford & von Noorden, 1980) and illustrated in figure 7. Moreover, these binocular cells do not recover function within 90 days following dissociation (filled triangles) or even with a period of normal binocular vision as long as three years (Crawford, et al., 1984) as indicated by the open triangles of figure 7.

We wanted to test the hypothesis that binocular cells are necessary for stereopsis, and therefore, we carried out a series of behavioral experiments on young monkeys which had had optical dissociation of the visual images for a period of 30 days beginning at 30 days of age. We employed two measures of binocular function. The first measure was the superiority of threshold detection when the animal was permitted to view a sinewave grating with both eyes, compared to viewing such a grating monocularly (Campbell & Green, 1965). The second test was to measure the ability of normal and

Figure 7. The percentage of binocular neurons in the visual cortices of normal and optically dissociated monkeys. The square on the ordinate shows adult monkeys to have 77% binocular cells, the open circle shows 30 day old monkeys to have 65% binocular cells, and the solid circles indicates that 30-120 days of age animals normally have about 72% binocular cells. Viewing the world through prisms reduces this percentage in a near-linear fashion, until only 10% binocular cells remain after 57 days. The filled triangles indicate the percentages of binocular neurons found after 30, 60, or 90 days of normal experience following the prism rearing between 30-60 days of age. The open triangles show the percentages of binocular neurons found 3 years after 30 days of optical dissociation. Binocular neurons do not recover. The solid square indicates the binocular cells found in a monkey reared for 30 days inside a striped cylinder (von Noorden & Crawford, 1981). Rearing in a striped cylinder during strabismus does not preserve binocularity.

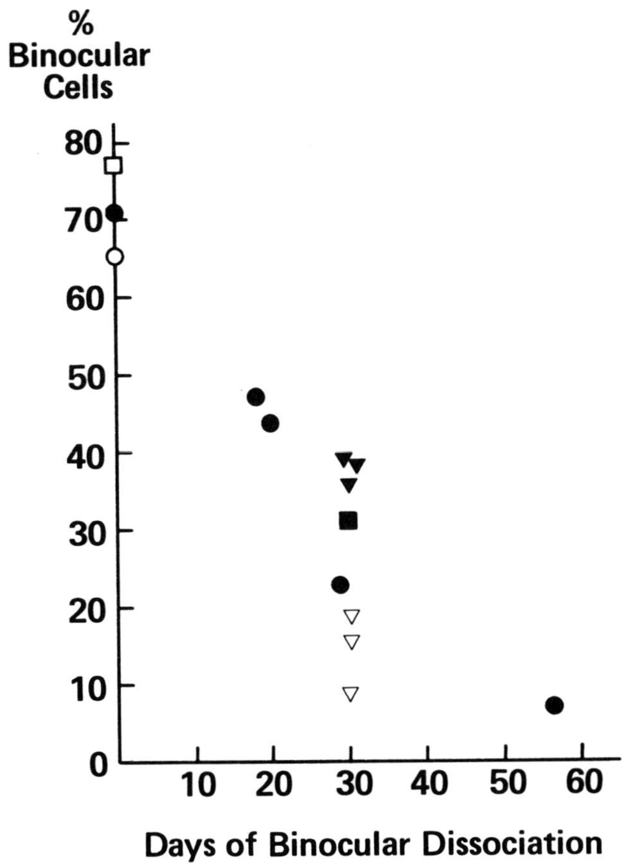

%
Binocular
Cells
80
70
60
50
40
30
20
10
0
10
20
30
40
50
60
Days of Binocular Dissociation

optically dissociated monkeys to detect dynamic random dot stereograms, and finally, we compared the performance of the monkey subjects with human subjects who had clinical histories of early strabismus (Crawford, et al., 1983).

Figure 8 illustrates examples of the results from these studies. Figure 8A shows the contrast sensitivity function of a normal monkey for viewing sine wave gratings as previously described. Three sets of data points are presented on each graph which represent contrast sensitivity of the monkey when using one eye, or the other eye, or when using both eyes in binocular viewing. It is to be noted that a normal monkey subject (L38) had comparable sensitivity in the right and the left eye, but showed an advantage in sensitivity when both eyes were used for the detection task (open squares). Like human subjects, it is clear that the normal monkey enjoys a superior sensitivity when using both eyes as compared to using either eye alone. Compare this with the curves of 8B where the optically dissociated experimental subject L51 shows no particular improvement in sensitivity in detecting the gratings by bincoular viewing, again indicated by the open squares. In summary the binocular dissociated subject (L51) exhibits no amblyopia as a result of having worn prisms before his eyes for 30 days early in life, and yet there is no evidence of binocular advantage in sensitivity as compared to the normal monkey subject (L38).

The monkeys were then trained to detect the brief disappearance of grating images embedded in anaglyphic dynamic random dot stereograms presented upon a color television screen at a 1 meter viewing distance. To the normal human observer, the stimulus was perceived as a clear square wave grating standing out in depth. Examples of the results of these experiments are shown at the bottom of figure 8C. Normal control monkeys such as L38 (shown on the left

Figure 8. The effects of optical dissociation and strabismus upon binocular functions in monkeys and children. A: the normal contrast sensitivity functions for viewing with the right or left eyes (circles) or when both eyes are used (squares). The squares represent binocular summation, and higher sensitivity. B: The same as A, but for an optically dissociated monkey, showing no particular binocular summation. C: The ability of the normal and the optically dissociated monkey to detect dynamic random-dot stereograms, compared to similar detection performance of a normal (M.M.), and two human observers who had clinical histories of early strabismus. Striped bar indicates detection performance with a luminance cue, the stippled bar indicates detection level for the stereograms alone.

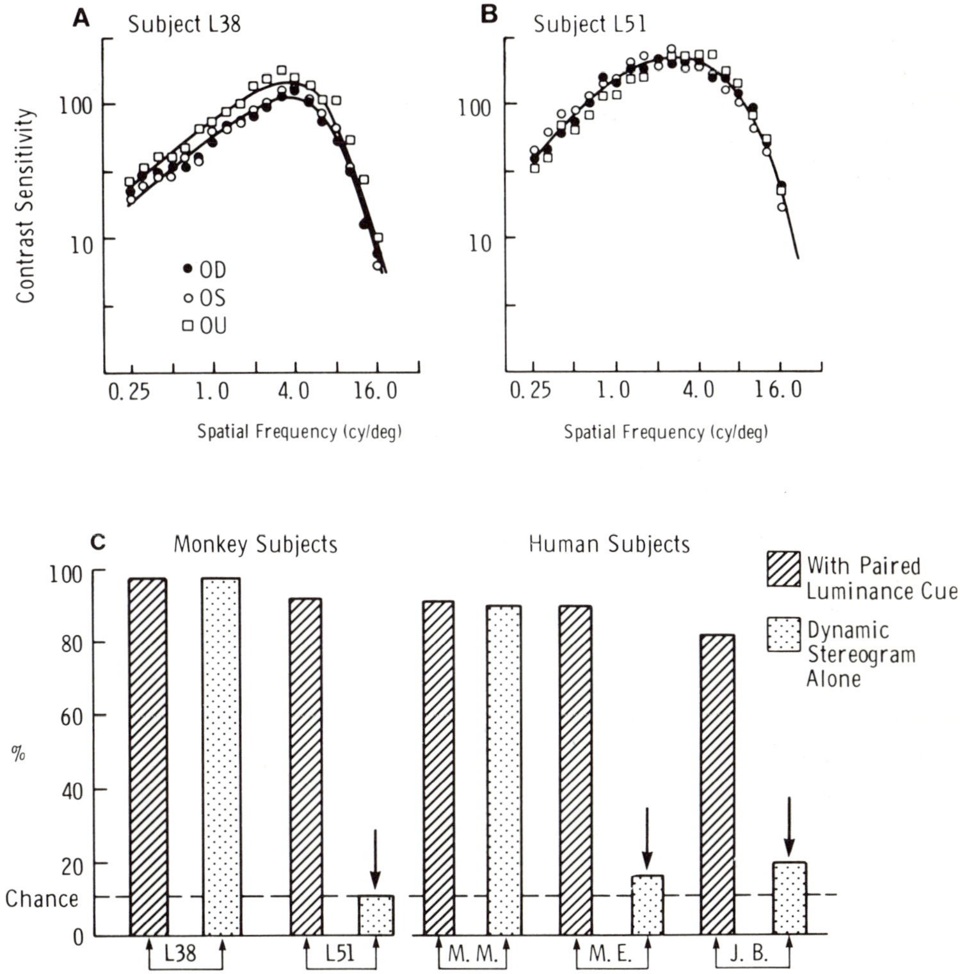
A
Subject L38
Contrast Sensitivity
100
10
OD
OS
OU
0.25
1.0
4.0
16.0
Spatial Frequency (cy/deg)
B
Subject L51
100
10
0.25
1.0
4.0
16.0
Spatial Frequency (cy/deg)
C
Monkey Subjects
Human Subjects
With Paired Luminance Cue
Dynamic Stereogram Alone
100
80
60
%
40
20
Chance
0
L38
L51
M. M.
M. E.
J. B.

of the panel) could detect the appearance or disappearance of the stimulus figure in the random dot display with high proficiency (>95%) either when given a luminance cue or when the luminance cue was removed and the animal had to sustain his detection performance based only upon the dynamic stereogram image alone. This normal monkey could clearly see the stereogram image, and therefore had stereopsis. Compare this excellent performance with that of L51 who had an equally high performance for detection when the stereo image was paired with a luminance cue, but his performance fell to chance level when he had to rely upon the stereo cue alone. Clearly, this monkey who had optically induced strabismus early in life and no amblyopia, gave every indication of being stereoblind. Comparable results have been obtained on other monkeys treated in the same manner.

Children with similar histories of congenital strabismus, surgical realignment, and successful therapy to avoid amblyopia were tested in the same manner as the monkeys and the results are shown in figure 8C. Like the normal monkey, the normal 8-year old girl (M.M.) easily detected the stereogram and verbally gave an accurate description of it. No doubt she could readily work the task, see the stereograms, and had stereopsis. However, two other teenage girls (M.E. & J.B.) who had 20/20 vision at the time of testing, scarcely did better than guessing on the task when the luminance cue was removed. Each girl had been diagnosed as having congenital intermittant strabismus, becoming manifest later, corrective surgery followed by successful therapy to produce normal acuity in each eye. They were totally oblivious to the random dot image, and as the monkeys, were clearly stereoblind.

What remained to tighten the link between cortical binocularity and stereopsis was the demonstration that these stereoblind monkeys had significantly fewer binocular neurons in their striate cortices. And indeed they did, having only 22% binocular neurons as compared to a robust 81% found in normal monkeys (Crawford, et al., 1984). Figure 9 illustrates these average eye-dominance profile differences, for both striate (V1) and pre-striate (V2) cortex. When the average eye-dominance histogram for 372 neurons from four normal monkeys (dark bars) is compared with that for 276 neurons from the stereoblind monkeys, the differences are dramatic.

In summary, these experiments point out the similarity of sensitivity of the two primate visual systems to abnormal early

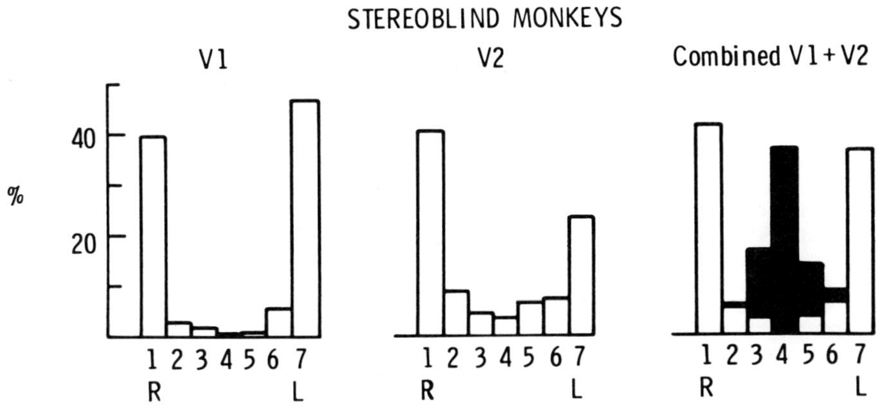

Figure 9. The average eye-dominance profiles for neurons from three optically dissociated stereoblind monkeys. Profiles are shown for 276 neurons from striate and 108 neurons from prestriate cortex. The dark histogram superimposed upon the combined histogram, comes from 485 neurons recorded from four normal monkeys. It is clear that the stereoblind monkeys have few binocular cells.

visual experience and emphasize the importance of early clinical diagnosis and corrective measures for congenital visual anomalies.

REFERENCES

Baker, F.H., Grigg, P. and von Noorden, G.K. (1974) Effects of visual deprivation and strabismus on the response of neurons in the visual cortex of the monkey, including studies on the striate and prestriate cortex in the normal animal. Brain Res. 66, 165-208.

Barlow, H.B., C. Blakemore and J.D. Pettigrew (1967) The neural mechanism of binocular depth discrimination. J. Physiol., 193, 327-342.

Bishop, P.O. (1973) Neurophysiology of binocular single vision and stereopsis. In: The Handbook of Sensory Physiology, Vol. VII/3, Ed. by R. Jung, pp. 255-305, Springer-Verlag, Heidelberg.

Campbell, F.W. and Green, D.G. (1965a) Optical and retinal factors affecting visual resolution. J. Physiol. (Lond.) 181, 576-593.

Campbell, F.W. and Green, D.G. (1965b) Monocular versus binocular visual acuity. Nature, 208, 191-192.

Crawford, M.L.J. (1978) The visual deprivation syndrome. Trans. Ophth. A.A.O.O. 85:465-477.

Crawford, M.L.J. and von Noorden, G.K. (1979) Concomitant strabismus and cortical eye dominance in young rhesus monkeys. Trans. Ophthalmol. Soc. UK, 99, 369-174.

Crawford, M.L.J. and von Noorden, G.K. (1980) Optically induced commitant strabismus in monkeys. Invest. Opthalmol. Vis. Sci., 19, 1105-1109.

Crawford, M.L.J., von Noorden, G.K., Meharg, L.S., Rhodes, J.W., Harwerth, R.S., Smith, E.L. and Miller, D.D. (1983) Binocular neurons and binocular function in monkeys and children. Invest. Ophthalmol. Vis. Sci., 24, 491-495.

Crawford, M.L.J., Smith, E.L. III, Harwerth, R.S., and G.K. von Noorden (1984) Stereoblind monkeys have few binocular cortical neurons. Invest. Ophthalmol. Vis. Sci., 25, 779-781.

Harwerth, R.S. and Sperling, H.G. (1975) Effects of intense visual radiation on the increment-threshold spectral sensitivity of the rhesus monkey eye. Vision Res. 15, 1193-1204.

Harwerth, R.S. (1982) Behavioral studies of amblyopia in monkeys. Am. J. Optom. Physiol. Optics, 59, 535-555.

Harwerth, R.S., Crawford, M.L.J., Smith, E.L. and Boltz, R.L. (1981) Behavioral studies of stimulus deprivation amblyopia in monkeys. Vision Res., 21, 779-789.

Harwerth, R.S., Smith, E.L., Crawford, M.L.J. and von Noorden, G.K. (1982) Increment-threshold spectral sensitivity in monkeys with experimental amblyopia. Invest. Ophthalmol. Vis. Sci. (suppl.), 22, 89.

Harwerth, R.W., Smith, E.L., Boltz, R.L., Crawford, M.L.J., and von Noorden, G.K. (1983) Behavioral studies on the effect of abnormal early visual experience in monkeys: Spatial modulation sensitivity. Vision Res., 23, 1501-1510.

Harwerth, R.S., Smith, E.L., Boltz, R.L., Crawford, M.L.J. and von Noorden, G.K. (1983) Behavioral studies on the effects .cp3 of abnormal early visual experience: Temporal modulation sensitivity. Vision Res., 23, 1511-1517.

Harwerth, R.S., Smith, E.L. III, Crawford, M.L.J. and von Noorden, G.K. (1984) Effects of enucleation of the nondeprived eye on stimulus deprivation amblyopia in monkeys. Invest. Ophthalmol. Vis. Sci., 25, 10-18.

Hubel, D.H. and T.N. Wiesel (1962) Receptive fields, binocular interaction and functional architecture in the cat's visual cortex. J. Physiol., 160, 106-154.

Smith, E.L. III, R.S. Harwerth & M.L.J. Crawford (1985). Spatial contrast sensitivity deficits in monkeys produced by optically induced anisometropia. Invest. Ophthal. & Vis. Sci. 126, 330-342

Sperling, H.G. and Harwerth, R.S. (1971) Red-green cone interactions in increment-threshold spectral sensitivity of primates. Science, 172, 180-184.

von Noorden, G.K. and Dowling, J.E. (1970) Experimental amblyopia in monkeys II. Behavioral studies in strabismus amblyopia, Arch. Ophthal., 84, 215-220.

von Noorden, G.K., Dowling, J.W. and Ferguson, D.C. (1970) Experimental amblyopia in monkeys I. Behavioral studies of stimulus deprivation amblyopia. Archs. Ophthal., 84, 206-214.

von Noorden, G.K. and M.L.J. Crawford. (1981) Failure to preserve cortical binocularity in strabismic monkeys raised in unidirectional visual environment. Invest. Ophthal. & Vis. Sci. 20:665-670.

Visual System Plasticity in Human Amblyopia

Kenneth J. Ciuffreda

If the visual system is exposed to a period of abnormal visual experience resulting in diplopic and/or degraded retinal images, especially early in life when visual system plasticity is maximum, changes in the neural pathways subserving vision can be demonstrated anatomically, physiologically, and psychophysically (Movshon and Van Sluyters, 1981; Sherman and Spear, 1982; Harwerth, 1982; Levi and Harwerth, 1982; Odom, 1983; Rauschecker, 1984; Garey, 1984; Mitchell and Timney, 1984). In lower animals such as the cat and monkey, as well as in man, the visual system effects of this abnormal visual experience include such abnormalities as reduction in lateral geniculate cell size (Wiesel and Hubel, 1963; Garey and Vital-Durand, 1981; Crewther et al, 1983; von Noorden et al, 1983; Tieman et al, 1984; Jones et al, 1984) and geniculo-cortical synapses (Tieman, 1984), reduction in number of binocular cortical neurons (Hubel and Wiesel, 1965; von Noorden and Crawford, 1979), decreased cortical neuronal responsivity (Hubel and Wiesel, 1965; von Noorden and Crawford, 1979) (in some cases due to inhibition from the fellow dominant eye; Blakemore et al, 1982; Hoffmann and Lippert, 1982), reduced visual resolution (Jones et al, 1984), and impaired sensory-motor coordination (Dews and Wiesel, 1970), with the degree of disruption of normal visual system development dependent to an extent on the severity, length, and timing of the abnormal visual experience. (Mitchell and Timney, 1984). For example, if a cat is monocularly occluded from the time of eye opening (≈10 days of age) through the third month of life, all of the above-mentioned deficits will be manifested to a marked degree; if the same is done but only at the peak of the so-called "critical" or "sensitive" period of visual development for a day or so, the overall effects are less pronounced, although cortical binocularity is still severely affected; the effects would be much less if the day of occlusion occurred at a non-peak period; and if any occlusion occurs beyond this sensitive period, the effects, if any, are generally minimal. While such a precise timetable has not yet been determined in man for all vision functions (Banks et al, 1975; Vaegan and Taylor, 1979; Jacobson et al, 1981), little if any amblyopia will develop if a period of abnormal visual experience occurs after the age of about 6 years (von Noorden and Crawford, 1979).

While defining the parameters required for normal visual system development is indeed important to both the basic scientist and clinician involved in this area, a second area which is equally important, somewhat neglected, and perhaps even more directly relevant to the clinician intimately involved in amblyopia therapy is that of physiological reversibility of these effects of early abnormal vision experience, e.g., how well can we regain vision function in a patient once it has already been severely compromised? Since preventive eye care and careful monitoring of patients who are "at risk" of developing vision problems are relatively new concepts in American health care, many more patients are presently examined who already exhibit the visual abnormality than have a high probability of yet developing the anomalous condition. Thus, the question of reversibility rather than evolution of the ocular condition is of immediate concern to the clinician faced with such a patient, although both issues are of equal importance, especially as long-term public health problems (Flom and Kerr, 1965; Allen, 1967; Tommila and Tarkkanen, 1981; Hillis et al, 1983).

Over the past decade or so, evidence in lower animals such as the cat and monkey has begun to accumulate regarding physiological reversal of these visual system effects, primarily when both the period of abnormal visual experience and the reversal maneuver occur during the sensitive period for vision development (Blakemore and Van Sluyters, 1971; Van Sluyters, 1978; Blakemore et al, 1981; Garey and Vital-Durand, 1981; Crewther et al, 1981; Swindale et al, 1981; Levitt and Van Sluyters, 1982; Crewther et al, 1983; Malach et al, 1984), although post-critical period reversal and plasticity has also been demonstrated (Creutzfeldt and Heggelund, 1975; Kratz et al, 1976; Hoffmann and Cynader, 1977; Singer et al, 1982). While such results leading to animal models of visual system development and plasticity have helped both the basic scientist and clinician, extension to include data and models based on careful measurements in humans has only recently been undertaken and: (1) allows for determination of species generalization of visual system plasticity, and (2) provides the clinician with an age-related data-base useful in arriving at a more accurate prognosis for recovery of vision function in patients with amblyopia. Clinicians have been treating patients manifesting the effects of such abnormal visual experience, which produces overall reduced and anomalous sensory function as manifested clinically by the presence of amblyopia, eccentric fixation, and suppression (See Appendix for glossary) for many years (actually many centuries; Duke-Elder and Wybar, 1973). However, there has been relatively little careful documentation, especially incorporating use of objective recording techniques and/or careful psychophysical procedures. In this chapter, I will review, primarily by case illustration, the results determined in my laboratory over the past several years, with additional supporting documentation from other laboratory and clinical investigations.

CASE PRESENTATIONS

CASE ONE
(Selenow and Ciuffreda, 1983)
Exotropic Amblyope With High Unilateral Myopia (Age 6 1/2 yrs)

HISTORY

A 6 1/2 year old male was first examined in our clinic in October 1980. His initial vision examination was at 5 years of age, at which time a diagnois of amblyopia was made. He was prescribed spectacles for constant wear and direct occlusion 1 hr per day. Patching was discontinued after 2 weeks due to noncompliance. The mother sought several additional professional opinions and, in each case, was told that nothing could be done to improve vision in the amblyopic eye. Thus, wearing of the spectacles was also discontinued.

The patient's general health history revealed several significant findings. Delivery was 1 1/2 months late, and pneumonia occurred on the second day of life. Also, the mother noted partial ptosis of the left eye several weeks after birth and a left exotropia at about 1 1/2 years of age. All post-natal developmental milestones were within expected age limits.

VISION EXAMINATION FINDINGS

There was a 3 mm lid ptosis of the left eye. Visuoscopy showed steady, central fixation in the right eye; the left eye showed unsteady, eccentric fixation approximately 75% of the time with a 1 degree superior-nasal bias. Subjective refraction was plano in the right eye (20/20) and -8.00 diopters in the left eye (20/200).

A-scan ultrasound showed axial lengths of 22.7 and 24.9 mm in the right and left eyes, respectively. Cover test revealed a constant, comitant, left exotropia of 14Δ (prism diopters) at 6M and 12Δ at 40 cm. Suppression of the left eye was noted at all distances with the Worth-4-Dot Test. There was no Randot stereoscopic appreciation. (See Table I)

TRAINING

There were three phases to the training. Phase one involved amblyopia therapy exclusively. This included monocular eye movement, accommodation, and eye-hand training procedures (Schapero, 1971; Griffin, 1976). In addition, the full spectacle correction was prescribed for full-time wear. Phase two emphasized development and training of binocular vision function with emphasis on fusion training. Phase three involved the fitting of a soft contact lens to the amblyopic eye. This was done to decrease the variable prismatic demand produced by the high minus spectacle lens when training fusion in nonprimary directions of gaze, as well as to reduce the unequal retinal-image sizes which were reported whenever diplopia was elicited. Antisuppression training performed simultaneously with eye movement, accommodation, and eye-hand coordination exercises during bifoveal fixation was also begun. Home training was performed during all three phases for 30 min a day. It consisted of procedures similar to those done in the clinic. In addition, direct occlusion was used 3 hr each day.

RESULTS

The improvements in monocular and binocular vision functions during the course of orthoptic therapy are presented in Fig. 1. Visual acuity increased throughout therapy, with this gain being especially large and rapid during phase one. Similar trends, but with more variability, were found for accommodation. Central fixation was rapidly achieved in the amblyopic eye, with steadiness consistently improving during the course of therapy. Reduction of the deviation occurred at both distance and near. Once the contact lens was fitted, stereoacuity improved rapidly. After termination of treatment, there was a period during which stereopsis exhibited some regression. However, stereoacuity returned to normal when measured 3 months after the end of treatment. The pre- and post-treatment clinical findings are summarized in Table 1. All clinical and laboratory findings remained unchanged when retested one year later, demonstrating good long-term maintenance of the substantial initial vision function improvements.

Monocular fixation with the amblyopic eye before and after orthoptic therapy is shown in Fig. 2. There was marked reduction in amplitude (2 deg pre- and less than 1 deg post-training) of saccadic intrusions (Ciuffreda et al, 1979a), as well as absence of intermittently increased drift (Ciuffreda et al, 1980b), after therapy. Monocular fixation with the dominant eye was within normal limits at each test session.

Threshold contrast sensitivity results taken 3 months after the end of training are presented in Fig. 3 for each eye (no pre-therapy measures were taken). Contrast sensitivity was normal in the dominant eye. Contrast sensitivity in the nondominant eye showed slight depression at the middle and high spatial frequencies.

TABLE 1. COMPARISION OF PRE- AND POST-TREATMENT CLINICAL FINDINGS IN THE AMBLYOPIC EYE (CASE 1)

Test	Pre-treatment	Post-treatment
Visual acuity (Snellen single line)	20/200	20/30
Visual acuity (S-chart)	20/109	20/32
Accommodative amplitude (push-up)	8.0 D	20 D
Accommodative amplitude (minus lens)	4.25 D	13 D
Visuoscopy	1 deg superior nasal unsteady	Central steady
Maximum fusional ability (6 M)	None	18Δ
Stereoscopic threshold (Wirt)	400 sec arc	40 sec arc
Stereoscopic threshold (Randot)	None	20 sec arc
Cover test (6 M)	14Δ left exotropia	7Δ exophoria
Cover Test (40 cm)	12Δ left exotropia	4Δ exophoria
Worth-4-Dot (40 cm and 6 M)	LE suppression	Fusion

CASE TWO
(Nathanson and Ciuffreda, 1982; Hokoda and Ciuffreda, in press)
Small-Angle Esotropic Amblyope (Age 9; 11 yrs)

HISTORY

The patient was initially examined in our vision training clinic in February 1978 at age 9 years. There was no previous history of eye surgery, orthoptic therapy, or correction of refractive error; however, his grandfather, aunt, and sister have strabismus. He had failed a school screening and was suspected of having a "lazy eye." The parents and patient wished to improve vision in the amblyopic eye.

VISION EXAMINATION FINDINGS

Uncorrected visual acuity (whole chart Snellen) was 20/20 in the dominant right eye and 20/200 in the amblyopic left eye. Refractive error was: right eye, plano; left eye, +0.25/-0.25 X 180; incorporation of the refractive error correction did not improve visual acuity in the amblyopic eye. Visuoscopy revealed slightly unsteady, central fixation in the dominant eye and very steady, superior-nasal, eccentric fixation (3 deg) in the amblyopic eye. Cover test revealed a small-angle (6Δ), constant, left esotropia at distance and near. The Worth Four-Dot-Test indicated suppression in the left eye for fixation beyond 25 cm. Anomalous retinal correspondence was indicated by the striated lens test and the major amblyoscope. Examination of the anterior and posterior segment of the eyes revealed absence of ocular or neurological disease.

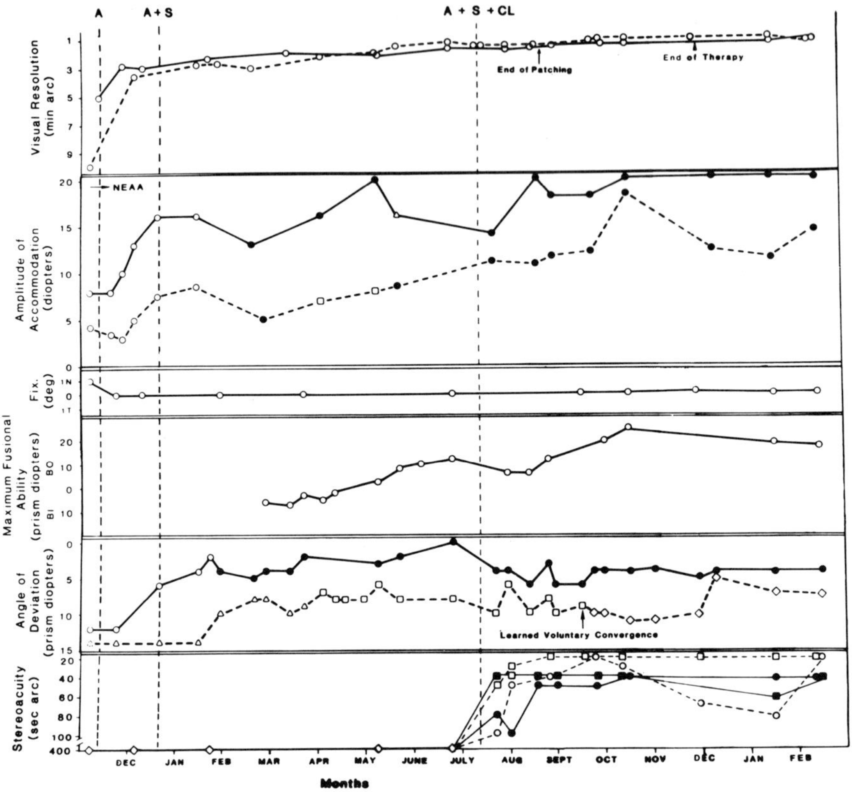

Fig. 1. Changes in vision function during the course of orthoptic therapy. A, amblyopia therapy; S, strabismus therapy; CL, contact lens correction. For visual acuity: dashed line is single-line Snellen and solid line is S-chart. For amplitude of accommodation: NEAA is normal eye amplitude of accommodation, while dashed line is minus lens and solid line is push-up accommodative amplitude in amblyopic eye; ○, 20/60 letter; □, 20/50 letter; ▲, 20/40 letter; and ●, 20/30 letter. For angle of deviation; ○, near tropia; near phoria; △, constant distance tropia; □, intermittent distance tropia; ◇, and distance phoria. For stereoacuity: ●, crossed disparity (Wirt); ■, uncrossed disparity (Wirt); O, crossed disparity (Randot); and □, uncrossed disparity (Randot).

TRAINING

The patient began formal vision training with weekly clinical sessions. Extensive home and clinic training included direct and indirect patching, eye-hand coordination exercises, and accommodative rock exercises (Schapero, 1971; Griffin, 1976). Training of central fixation was attempted by pleoptics and

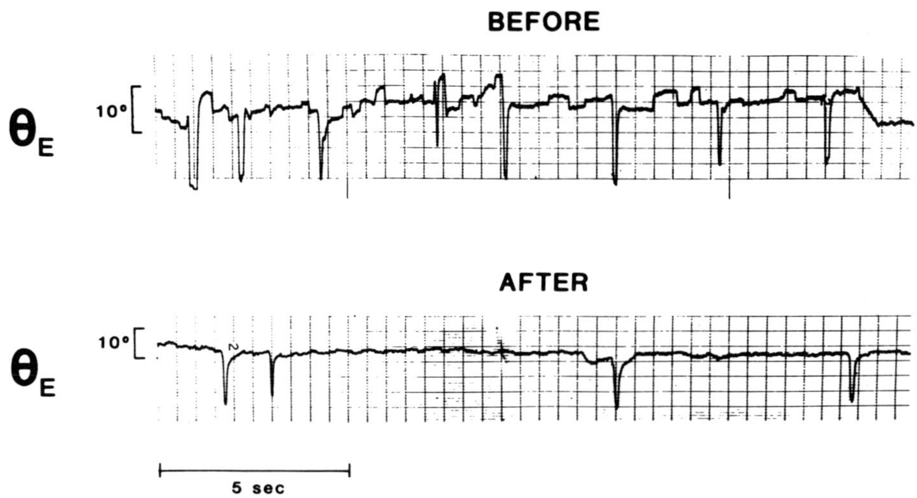

Fig. 2. Fixational eye movements before and after orthoptic therapy. Amblyopic eye. Monocular viewing. Post-training improvement is evident. Large, downward deflections represent blinks. Θ_E is eye position.

Haidinger brush fixation exercises. Visual acuity improved for 3 months, but then remained constant at 20/80 for 2 months. Fixation was still unsteady and eccentric (2 deg). It was decided to terminate therapy due to lack of further improvement of visual acuity and fixation ability.

The patient was referred to us 1 1/2 years later for a reevaluation. With the exception of visual acuity regressing to 20/120, all other findings remained as they were at the termination of therapy 1 1/2 years earlier. Because of the patient's cooperative nature and the recent dramatic results reported by Campbell and colleagues involving a physiologically-based grating treatment for amblyopia (Campbell et al, 1978), we decided to attempt intensive CAM treatment with this patient. A CAM-type grating device was used by the patient for daily home training. Five 7-min treatment periods were interspersed throughout the late afternoon and early evening hours during an 8-week period, totaling 30 hr of monocular grating stimulation to the amblyopic eye. No other treatment was used during this 2-month period. A few weeks after termination of CAM treatment, intensive conventional orthoptic treatment, including direct occlusion, accommodative rock exercises, eye-hand coordination training, and eye movement training (Schapero, 1971; Griffin, 1976) was reinstituted by us.

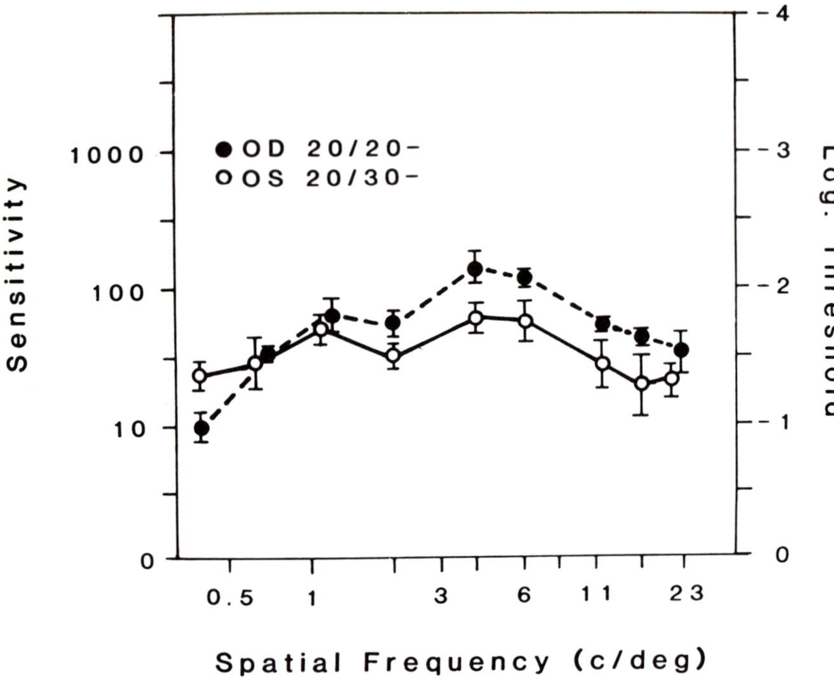

Fig. 3. Contrast sensitivity functions in dominant [●, (20/20+)] and nondominant [O, (20/30+)] eyes. Contrast sensitivity is normal in the dominant eye. Contrast sensitivity in the nondominant eye is normal only at very low spatial frequencies. Plotted are $\bar{x} \pm 1$ SD.

RESULTS

Fig. 4 presents the visual acuity results during the CAM treatment period. A small, steady improvement of visual acuity occurred as treatment progressed. The acuity improvement was approximately 13% Snell-Sterling, thus just exceeding our 10% criterion level indicating a significant change (Ciuffreda et al, 1980a). Correlation analysis indicated significant improvement of visual acuity over the sessions for values obtained before (rho=-0.8095, $p<0.05$) and after (rho=-0.9047, $p<0.01$) a laboratory CAM treatment. Visual acuity 2 weeks after termination of the CAM treatment (week 10) was similar to that measured at the last session (week 8), suggesting persistence of the small treatment effect (probably due to training of monocular fixation rather than to any special attributes of the stimulus per se; Ciuffreda et al, 1980) and lack of a significant consolidation-type effect (Freeman and Olson, 1979).

Figures 5 to 8 present the results obtained during the subsebquent 1 1/2 year period of intensive, conventional orthoptic therapy. S-chart visual acuity profiles (Flom et al, 1963) obtained at selected times during this period are presented in Fig. 5. Two trends were evident. First, visual acuity in the amblyopic eye

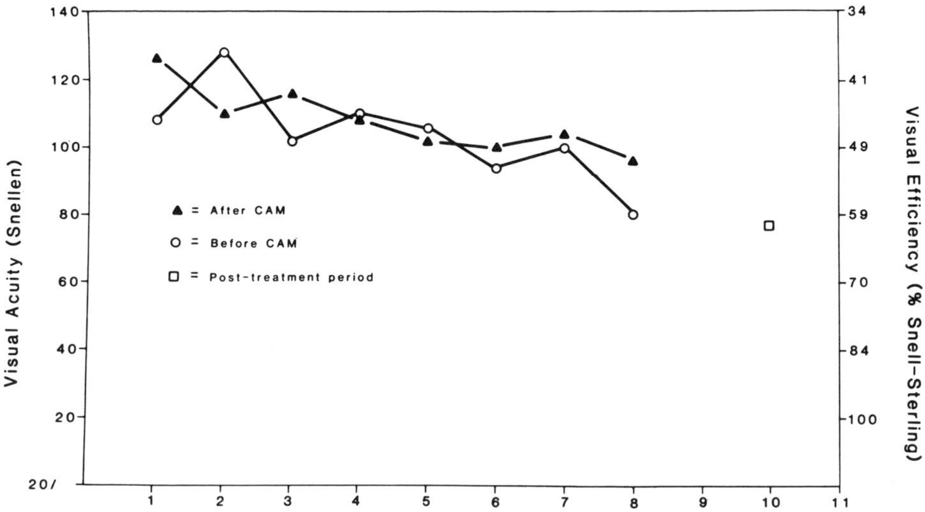

Fig. 4. Visual acuity and visual efficiency in the amblyopic eye as a function of weeks of CAM treatment.

gradually reached normal levels, although it was still slightly better in the fellow dominant eye. Second, the slope of the post-therapy (week 84) profile was normal and exceeded the more shallow slopes found at weeks 1 and 30. Changes in S-chart visual acuity, fixation, and interocular difference in accommodative amplitude during the course of therapy are presented in Fig. 6. There was a clear normalizing trend in each function, although rates of recovery differed. Accommodative amplitude improved the most rapidly. While changes in locus of fixation and visual acuity exhibited similar rapid changes during weeks 25 to 40, this relatively small change in fixation (± 1 deg) could not account for most of the visual acuity improvement (Borish, 1970). Vision function improvement persisted immediately following therapy (weeks 82 to 90) and remained unchanged one year later. Changes in the steady-state accommodative stimulus/response profiles are presented at selected times during the therapy period in Fig. 7. There was a clear normalizing trend. Slope in the amblyopic eye increased from 0.68 to 0.87 (it was 0.88 in the dominant eye), so that response profiles in the amblyopic eye post-therapy and the dominant eye were essentially superimposed. Lastly, the accommodative spatial frequency response profile (Ciuffreda and Hokoda, 1983; Hokoda and Ciuffreda, in press) in each eye was assessed post-therapy (Fig. 8). It is of interest that while visual acuity in each eye was now about equal (20/13 versus 20/17), and accommodation to square-wave type stimuli (i.e., Snellen letters) was normal in the (formerly) amblyopic eye (Fig. 7), static accommodative responses to simple high contrast (≈80%), sine-wave gratings were not. Response amplitude and spatial frequency response range were less in the amblyopic eye. Also, the contrast sensitivity function (Fig. 8, inset) still showed slight overall depression in the amblyopic eye.

Fig. 5a. Test chart (S-chart; Flom et al, 1963) used to assess visual acuity in the amblyopic eye accurately. At each letter size, there are 4 possible choices of letter orientation of the 8 letter E's presented. Clinically, one uses the 5 out of 8 correct response level as an estimate of visual acuity; in the research laboratory, probit analysis is used to calculate visual acuity.

CASE THREE
(Ciuffreda et al, 1979b)
Anisometropic Amblyope (Age 18 yrs)

HISTORY

The patient was an 18-year-old male who came to our clinic to obtain a replacement for his spectacles which has been lost 1 year earlier. He reported that vision has always been poor in the left eye and that due to a birth trauma, his left eye has been totally occluded during the first 2 days of life.

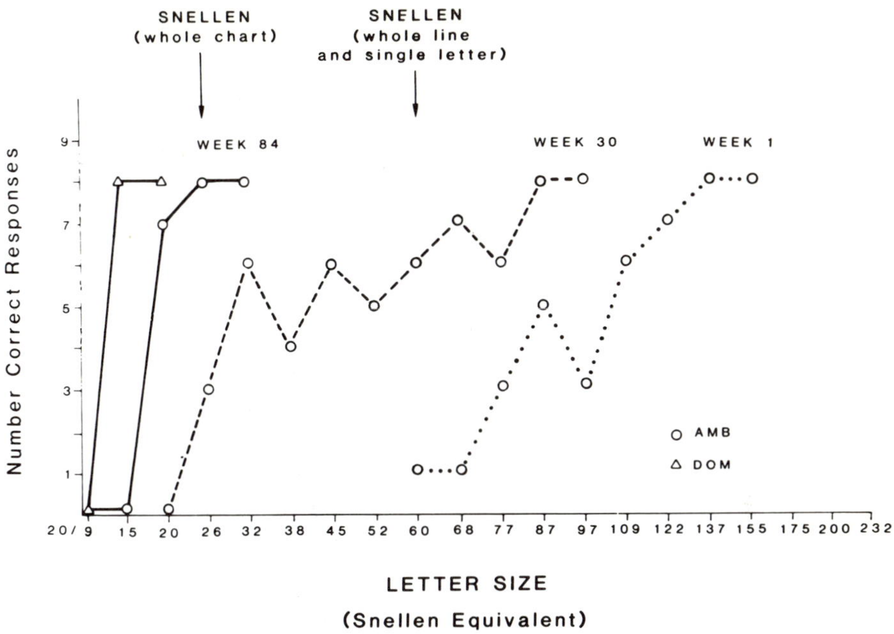

Fig. 5b. Visual acuity profiles (S-chart) at selected times during the course of orthoptic therapy. For weeks 30 and 84, the arrows indicate the Snellen visual acuity values.

VISION EXAMINATION FINDINGS

Refraction was +5.00 diopters in the left eye (20/230) and +3.00 diopters in the right eye (20/15). Cover tests clearly showed the absence of any strabismus. Initial visuoscopy indicated 10 prism diopters of temporal and 5 prism diopters of superior unsteady eccentric fixation. Haidinger brush and the afterimage transfer test indicated normal retinal correspondence. No ocular or neurological disease was detected. The patient indicated that he would be willing to undergo therapy in order to improve vision in his amblyopic eye.

TRAINING

The initial diagnostic tests and the first phase of orthoptic therapy were conducted prior to the author's involvement in the case. Orthoptic procedures were conventional (Schapero, 1971; Griffin, 1976) and included direct and inverse occlusion, transferred afterimage, eye-hand coordination activities, accommodation exercises, and in-office pleoptics. During this 6 month training period, eccentric fixation decreased to approximately 3 to 5 prism diopters temporal, and visual acuity was frequently as good as 20/50. However, neither measure was stable. The second phase of orthoptics in which new clinicians began their rotation lasted 10 months. The author became involved in the eye movement

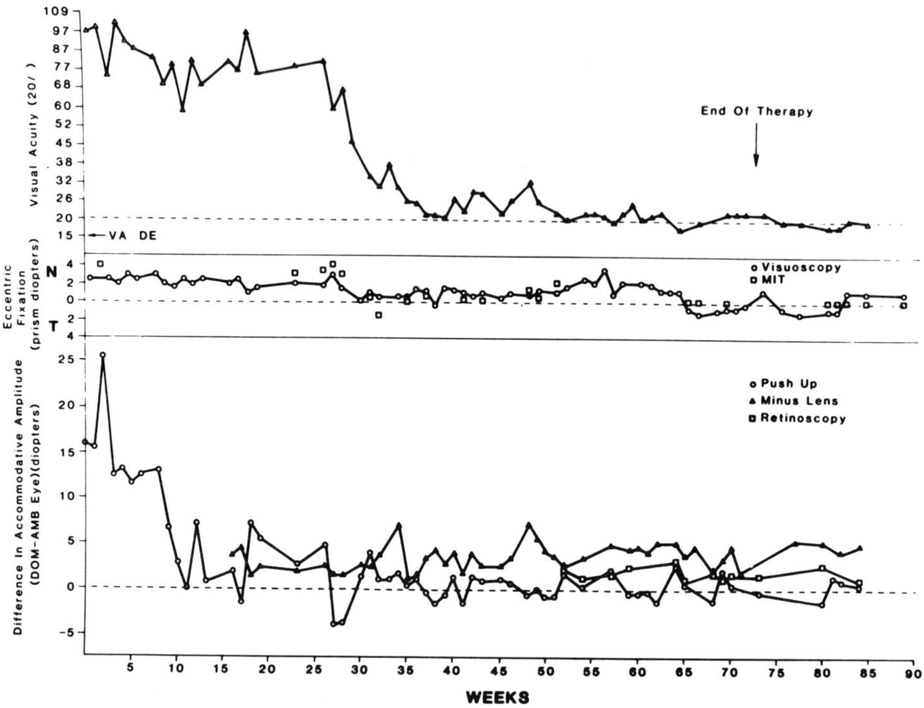

Fig. 6. Change in S-chart visual acuity, eccentric fixation, and interocular difference in accommodative amplitude during the course of orthoptic therapy.

testing, as well as some phases of clinical testing and training, during the last 8 months of this second phase. Orthoptic training procedures similar to those used in the first phase were instituted with the following additional techniques or modifications introduced at the indicated improved visual acuity levels; antisuppression training (20/90), fusion training (20/70), direct occlusion (20/50), and saccadic and pursuit tracking (20/25). Thus, as in most cases, a variety of therapeutic measures were implemented in order to increase the probability of a functional cure. Eye movements were recorded on six separate occasions during this second phase of training. Two months after 20/20 visual acuity was attained, the patient moved to another state and was no longer available for follow-up testing.

RESULTS

One of the most consistent findings of our study was the persistence of increased saccadic latencies in the amblyopic eye for central stimuli (Fig. 9). As fixation centralized and visual acuity normalized, saccadic latency remained approximately 100 msec longer for monocular tracking with the amblyopic eye than for either binocular tracking or monocular tracking with the dominant eye.

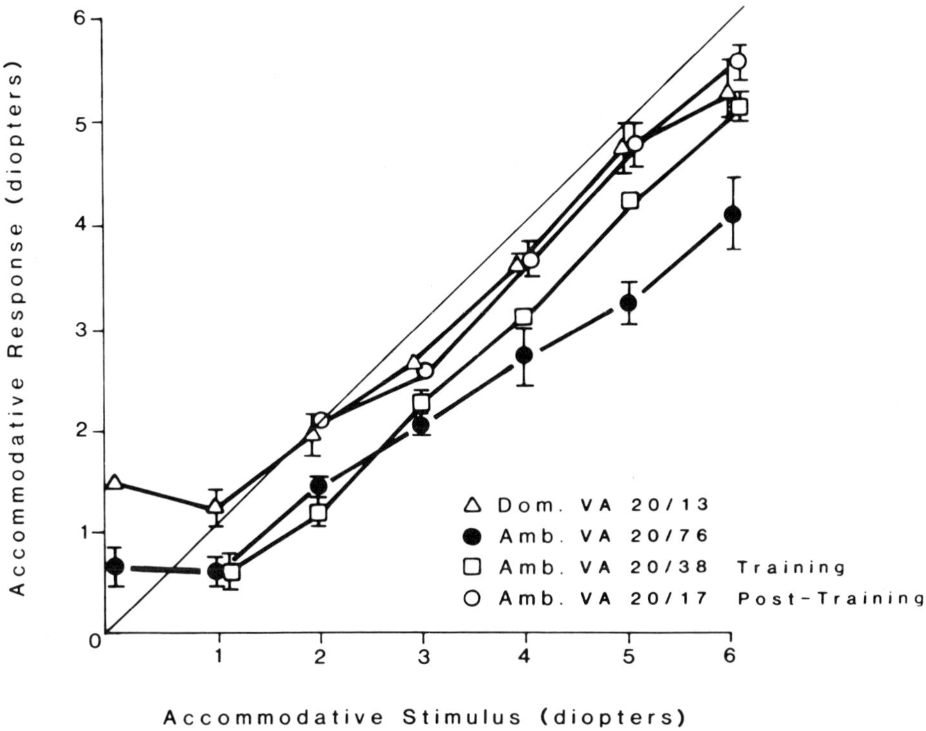

Fig. 7. Changes in steady-state accommodative stimulus/response profiles at selected times during the course of orthoptic therapy. Measured using a Hartinger coincidence optometer as the subject monocularly focused on letters of a reduced Snellen chart presented in a Badal optical system. Accommodative stimulus level was varied randomly. $\bar{x} \pm 1$ s.e.m. (n=6).

Fig 10 shows changes in fixational drift, eccentric fixation, and visual resolution during the course of the orthoptic treatment. In general, drift characteristics tended to normalize, and periods of steady fixation became more frequent as visual acuity and eccentric fixation normalized. Drift characteristics were similar for all five amplitudes of horizontal gaze tested (0, $\pm$ 2.5 deg, and $\pm$ 5.0 deg). Binocular fixation and monocular fixation with the dominant eye were within normal limits (<12 min arc drift amplitude, <20 min arc/sec drift velocity).

Two pursuit abnormalities, abnormal saccadic substitution and low pursuit gain, were observed during the course of treatment. Abnormal saccadic substitution during pursuit of small-amplitude target excursions (1 and 2 deg) was found at each test session; that is, rather than smooth movements, saccades having amplitudes generally two to five times larger than the peak-to-peak target amplitude were used to track the target (Ciuffreda et al, 1979c). Representative examples of this abnormal saccadic substitution recorded at three acuity levels for the same small-amplitude

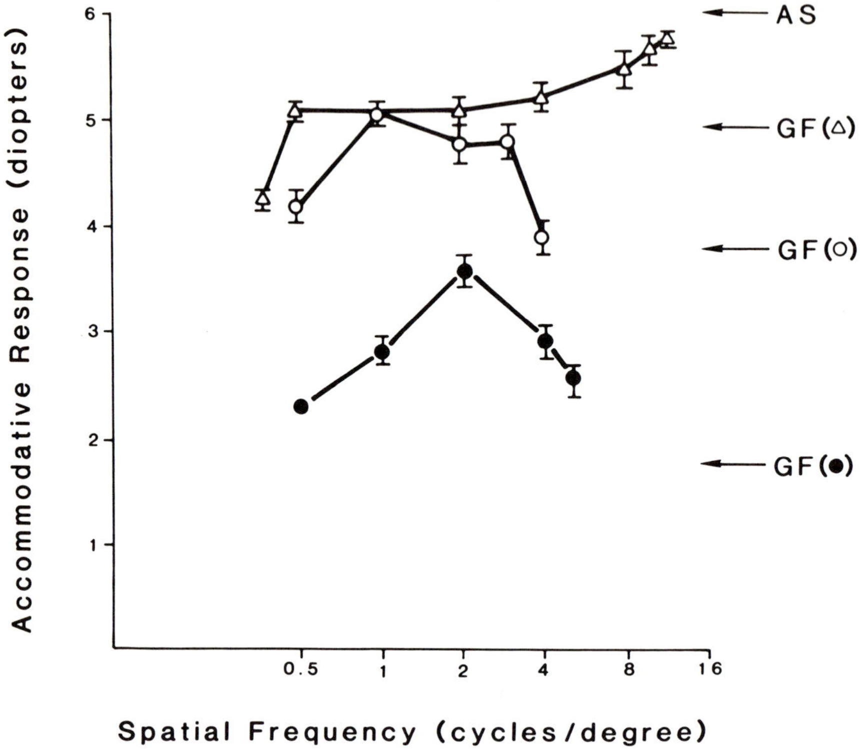

Fig. 8. Accommodative spatial frequency profiles at termination of orthoptic therapy in the amblyopic (o) and dominant (Δ) eye. Arrows indicate tonic accommodative levels, and AS indicates accommodative stimulus level. Measured using a Hartinger coincidence optometer as the subject monocularly focused on high contrast (80%), vertically-oriented sine-wave gratings whose spatial frequency was varied randomly. $\bar{x} \pm 1$ s.e.m. (n=6).

stimulus (1 deg., 3.75 deg/sec) are shown in Fig. 11. Large saccades, with little evidence of smooth movements, are prominent in each trace. Pursuit gain for tracking of targets having larger amplitudes (4 and 8 deg) was also of interest. When visual acuity was 20/110, long duration (up to 1 sec), low gain (0.1 to 0.2 range, 0.14 mean value) pursuit, followed by large saccades correcting for position errors, was observed. Low, variable pursuit gain (0.0 to 0.8 range, 0.45 mean value) continued at the 20/45 acuity level. However, once visual acuity reached 20/25, pursuit gain was generally higher and less variable (0.5 to 0.8 range, 0.60 mean value). Pursuit gain for monocular tracking with the dominant eye generally ranged from 0.7 to 0.95, with a mean value of 0.85 (within normal limits for our laboratory). Conjugate movements were always present during binocular pursuit tracking.

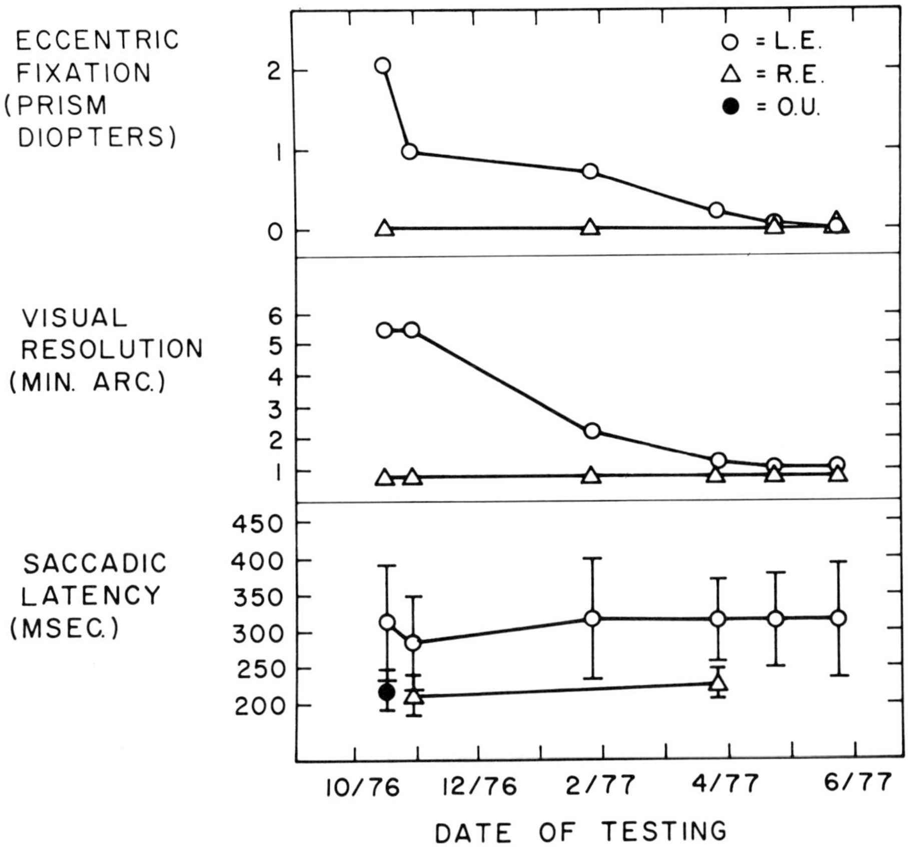

Fig. 9. Changes in eccentric fixation, visual resolution, and saccadic latency in each eye during the last 8 months of orthoptic therapy. Note centralization of fixation and normalization of visual resolution, but maintenance of increased saccadic latencies in the left amblyopic eye. Plotted are mean (and standard deviations for saccadic latency) of measures at each eye movement test session.

CASE FOUR
(Ciuffreda and Rumpf, in press)
Anisometropic Amblyope (Age 23 yrs)

HISTORY/VISION EXAMINATION FINDINGS AND THERAPY

The patient was a 23-year-old computer systems analyst who was seen by a private optometrist in July 1983. During the routine vision examination, the patient expressed a strong desire to improve vision in her amblyopic eye. She never had received spectacles, orthoptic therapy, or strabismus surgery; she remembered having a "lazy eye" since early childhood. Refraction was -5.75/-4.75 axis 165 (20/40) in the right amblyopic eye and -0.25/-1.50 axis 180

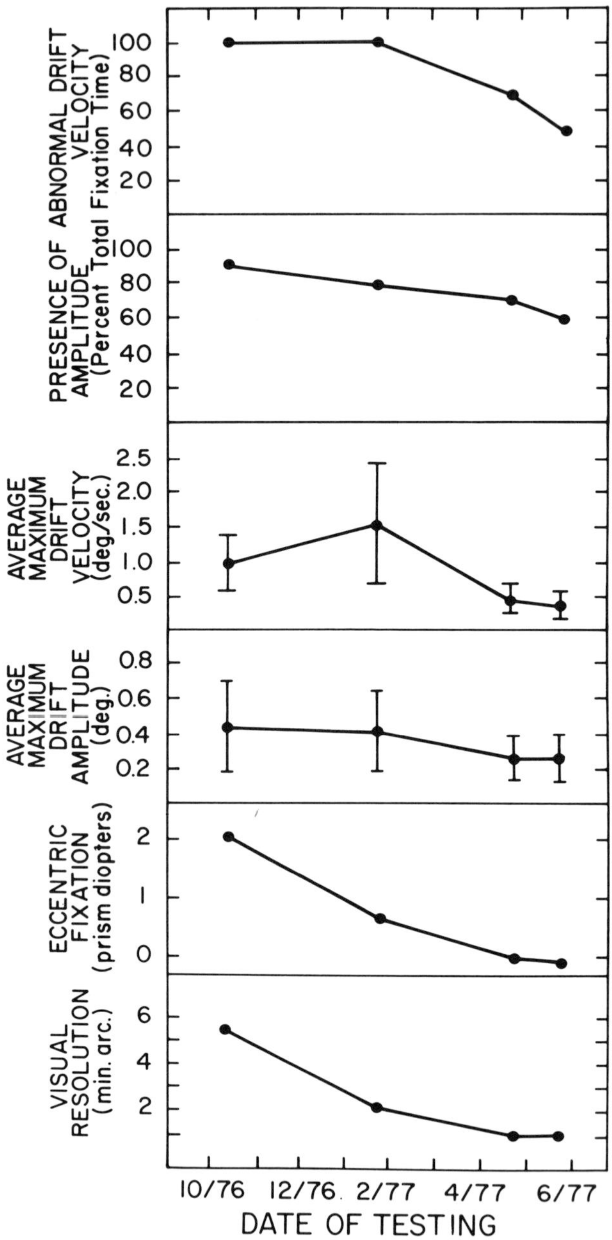

Fig. 10. Changes in the presence of abnormal drift velocity and amplitude, average maximum drift velocity and amplitude (with standard deviations), eccentric fixation, and visual resolution in the amblyopic eye during the last 8 months of orthoptic therapy. All clearly show normalizing trends.

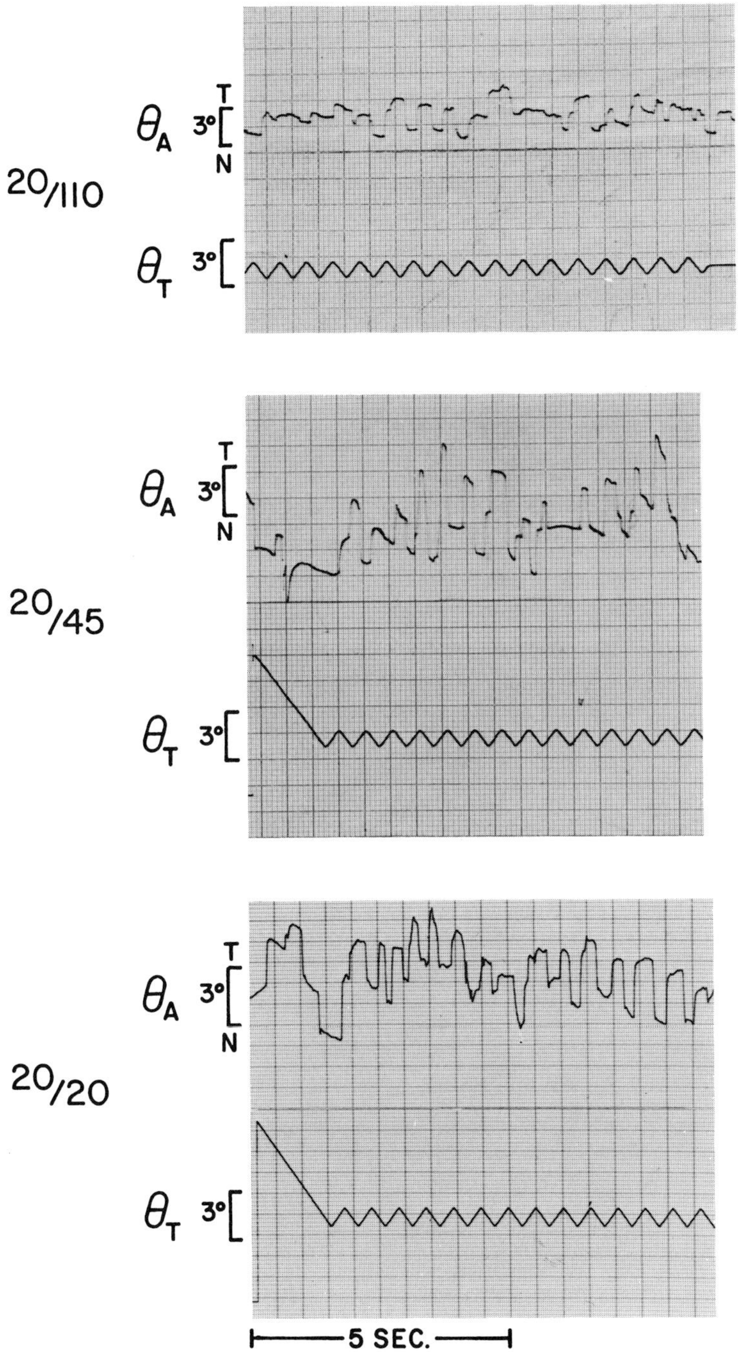

Fig. 11. Monocular pursuit with the amblyopic eye as a function of visual acuity level. In each pair of records, top trace is eye position, and bottom trace is the stimulus (1.0 deg amplitude, 3.75 deg/sec velocity, 1.88 Hz). Note persistence of abnormal saccadic substitution in spite of visual acuity improvement; this anomalous tracking response was most pronounced at the 20/45 level.

(20/20) in the fellow dominant eye. Cover test revealed absence of strabismus at distance and near. There was no evidence of ocular or neurological disease; general health status was good. Intense conventional orthoptic therapy (Schapero, 1971; Griffin, 1976), with emphasis on the accommodation system, in conjunction with part-time direct occlusion (3 hours/day), was instituted over a 5 month period.

RESULTS

The patient agreed to participate as a subject in an experiment involving the effect of contrast on accommodation. Briefly, the patient monocularly viewed either a vertical sine or square-wave grating whose contrast could be varied from 0.025 to 80.0%. Steady-state accommodation was monitored using a Hartinger coincidence optometer. The accommodation results before and after the 5 month period of orthoptic therapy to improve vision function in the amblyopic eye are presented in Fig. 12. Measurements were taken at four critical points: 80% contrast, the minimum contrast that still elicited an accommodative response significantly above the tonic level, the maximum contrast that elicited an accommodative response equivalent to the tonic level (i.e., -70 dB point), and the tonic accommodative level. Several important points deserve mention. First, the pre-therapy accommodative responses in the amblyopic eye were always less than in the fellow dominant eye. Second, the pre-therapy contrast cut-off value in the amblyopic eye was always smaller than in the dominant eye. Third, the post-therapy accommodative response level in the amblyopic eye now equalled or slightly exceeded that found in the fellow eye. And, fourth, the pre- versus post-therapy contrast cut-off value showed a trend toward normalization in the amblyopic eye. During this period, visual acuity improved from 20/40 to 20/25.

CASE FIVE
(Selenow and Ciuffreda, in press)
Small-Angle Esotropic Amblyope (Age 29 yrs)

HISTORY

A 29-year-old male college professor was referred to us by a local hospital with a diagnosis of functional amblyopia. We first saw the patient in our Amblyopia Clinic and Laboratory in February 1982. The patient was told that the amblyopia was discovered too late in life (age 11 years) to be improved by corrective lenses or therapy. The eye turn was first noticed by his parents at age 6 years. He had never been prescribed glasses or therapy. There was no history of trauma. Birth was by Caesarian delivery. Presently, a left esotropia was evident whenever the patient was tired or had moderate alcoholic intake, but there was no report of diplopia. The patient's goal was to improve vision in his amblyopic eye to allow a career change which requires at least 20/40 vision in the poorer eye. In addition, he wanted a more normally functioning "spare" eye. A complete amblyopia/strabismus workup was performed. Based on these findings, the prognosis for improvement of visual function in the amblyopic eye was made.

Fig. 12. Accommodation as a function of contrast in the dominant eye (▲), and in the amblyopic eye pre- (●) and post- (○) intensive orthoptic therapy, in an adult anisometropic amblyope.

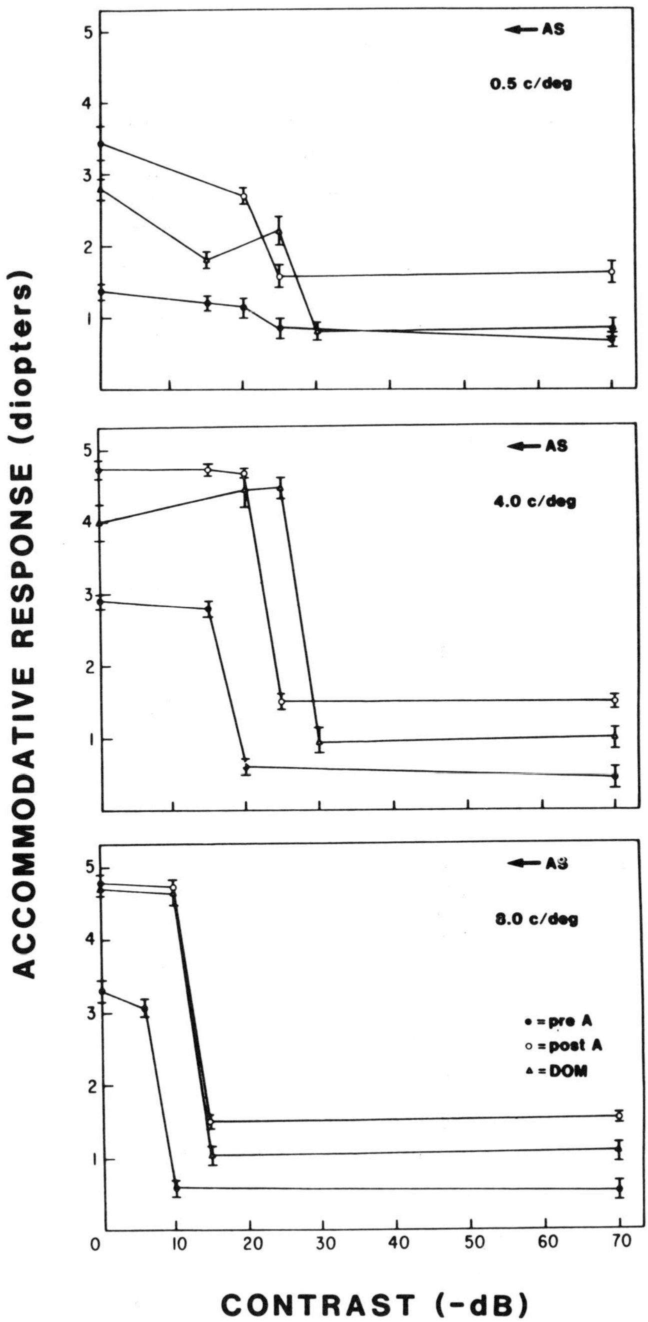
AS
0.5 c/deg
4.0 c/deg
8.0 c/deg
• = pre A
○ = post A
▲ = DOM
ACCOMMODATIVE RESPONSE (diopters)
CONTRAST (-dB)
0
10
20
30
40
50
60
70
1
2
3
4
5

VISION EXAMINATION FINDINGS

Snellen visual acuity (single-line) was 20/15 in the right eye and 20/150 in the left eye. S-chart visual acuity (Flom et al, 1963; Flom, 1966) was 20/18 in the right eye and 20/110 in the left eye. Visual acuity was the same with and without spectacle correction. Laser interferometric visual acuity was 20/20 in the right eye and 20/30 in the left eye. Retinoscopy was plano in the right eye and -0.25/-0.75 axis 55 in the left eye. Low-light visuoscopy showed steady central fixation in the right eye and three degrees of steady nasal eccentric fixation in the left eye. The cover test revealed a variable left esotropia of 6 to 14 prism diopters. The major amblyoscope revealed unharmonious retinal correspondence; the objective angle was 14 prism diopters, and the subjective angle was 7 prism diopters. There was no stereoacuity demonstrated on either the Titmus or Randot tests. Accommodative amplitude determined by the minus lens method was 7.25 and 6 diopters in the right and left eyes, respectively.

TREATMENT

Since one of the serious potential complications of occlusion therapy in an adult amblyope is the inducement of diplopia, our initial treatment consisted of teaching the patient to make divergent eye movements in response to diplopic images. Despite the fact that fixation was steady and eccentric, the amblyopia treatment plan was to begin with two hours of daily direct patching supplemented with one-half hour of oculomotor, accommodative, eye-hand coordination, and fixation training (Schapero, 1971; Griffin, 1976). In addition, thc patient received one hour of in-office testing and training every two weeks.

RESULTS

Several vision functions showed improvement throughout the course of treatment (Table 2). During the initial direct patching phase, both Snellen (20/150 to 20/40) and S-chart (20/118 to 20/62) visual acuities improved significantly (Figure 13). However, once indirect patching was instituted, both acuity values decreased. The second phase of direct patching again produced a marked improvement in Snellen acuity (20/70 to 20/30) and a more modest S-chart visual acuity improvement (20/99 to 20/75).

Although visual acuity improved during the initial phase of direct patching, eccentric fixation remained steady and eccentric. Only with subsequent indirect patching and use of foveal tagging could the anomalous fixation pattern be disrupted. After 16 weeks of indirect patching, initial fixation was still at the eccentric point; however, when asked to try and clear up the target, the patient immediately fixated with the fovea and reported the target to be clear and localized straight ahead. These observations were confirmed using objective recording procedures.

Electro-diagnostic testing was performed before and during (3 1/2 months later) the initial training period. Implicit times for the pattern-reversal cortical-evoked (≈105 msec) and electroretinographic (≈62 msec) responses were within normal limits and equal in each eye at each test session. However, response amplitudes in the amblyopic eye were reduced before training and

TABLE 2. COMPARISON OF PRE- AND POST-TREATMENT FINDINGS IN THE AMBLYOPIC EYE

Test	Pre-treatment	Post-treatment
Visual Acuity (Snellen single-line)	20/150	20/30
Visual Acuity (S Chart)	20/119	20/75
Visual Acuity (laser interferometry)	20/30	20/25
Visuoscopy	3 deg nasal steady	unsteady central
Accommodative Amplitude (minus lens)	6 D	9 D
Accommodative Spatial Frequency Profile	Overall depression	Normalized at middle & high spatial frequencies
Fixational Eye Movements	Small-amplitude jerk nystagmus	Reduced nystagmus ampli. & frequency
Contrast Sensitivity	Depression of middle & high spatial frequencies; low spatial frequency peak shift	Almost complete normalization
Visual Evoked Response	Reduced amplitude	Amplitude increase
Electro-retinogram	Reduced amplitude	Amplitude increase

tended to increase during training (See Table 2). Regarding the cortical-evoked responses, the ratio of response amplitude in the amblyopic to normal eye to 28 min arc checks was 0.42 pre-therapy and 0.85 3 1/2 months later; for 14 min arc checks, this ratio was 0.54 pre-therapy and 0.58 3 1/2 months later. Response amplitude of the electroretinogram increased by 40 percent (2.3 versus 3.2 microvolts) following the initial 3 1/2 months of intensive treatment. Further electro-diagnostic testing was not able to be performed for the duration of treatment.

The contrast sensitivity function was monitored before and during most of the training period (Figure 14). The pre-training response was markedly abnormal. It was reduced at all but the lowest (0.7 cycles/deg.) spatial frequency tested; further, the patient could not consistently detect a grating at spatial frequencies greater than 6 cycles/deg, even at the maximum contrast available. However, the response function improved considerably with training, by an amount even greater than has been documented in young children (Sjostrand, 1981). The peak shifted from 2 to 4 cycles/deg. and reduced sensitivity was only evident at spatial frequencies greater than 6 cycles/deg. Further, the amblyopic eye could now detect the grating at the highest spatial frequency available (23 cycles/deg), but required somewhat greater contrast than in the dominant eye. Once the patient could voluntarily fixate with either the fovea or the eccentric point, we were able to measure contrast sensitivity of the nasal versus temporal retina.

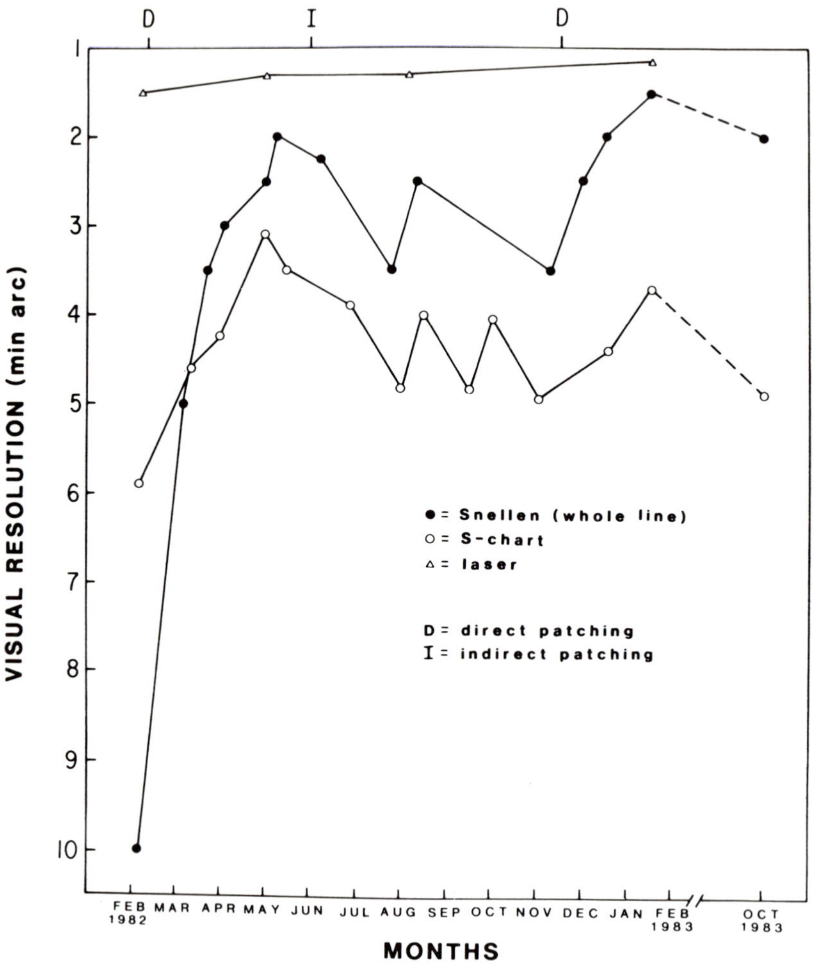

Fig. 13. Changes in visual resolution in the amblyopic eye during the course of treatment.

Nasal retinal contrast sensitivity showed a consistent depression at the middle and high spatial frequencies when compared to the temporal retina, which agreed with his subjective impression when viewing the Snellen chart. This result is consistent with previous reports of nasal-temporal threshold contrast asymmetries in eccentrically-fixating strabismic amblyopes (Thomas, 1978), and agrees with the notion of hemi-retinal suppression in strabismus.

Steady-state accommodation was assessed when focusing on two different types of stimulus (Fig 15). The first stimulus was a reduced Snellen chart presented in a Badal optical system at a range of dioptric levels (0 to 6 D), resulting in the classical

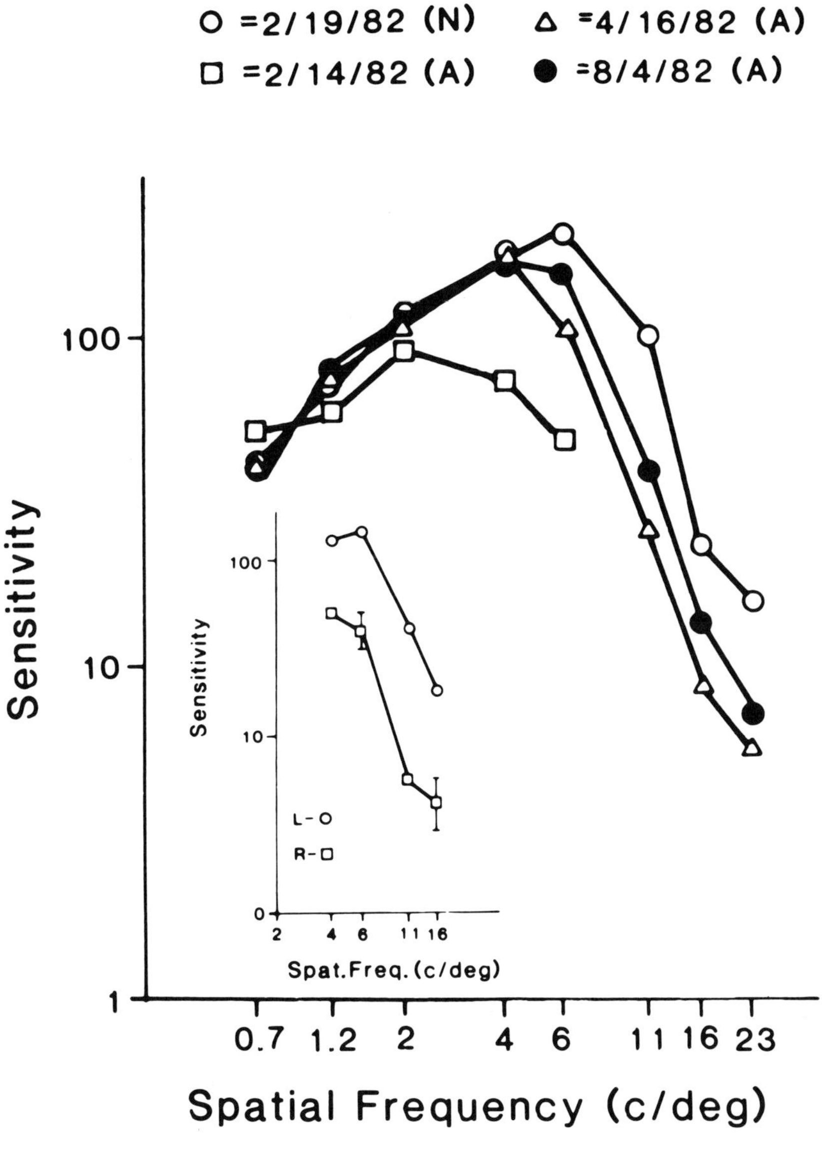

Fig. 14. Changes in contrast sensitivity during the course of treatment in the amblyopic eye (A); dominant eye (N) shown for comparsion. Inset shows nasal (R,□)/temporal (L, O) asymmetry of function in the amblyopic eye.

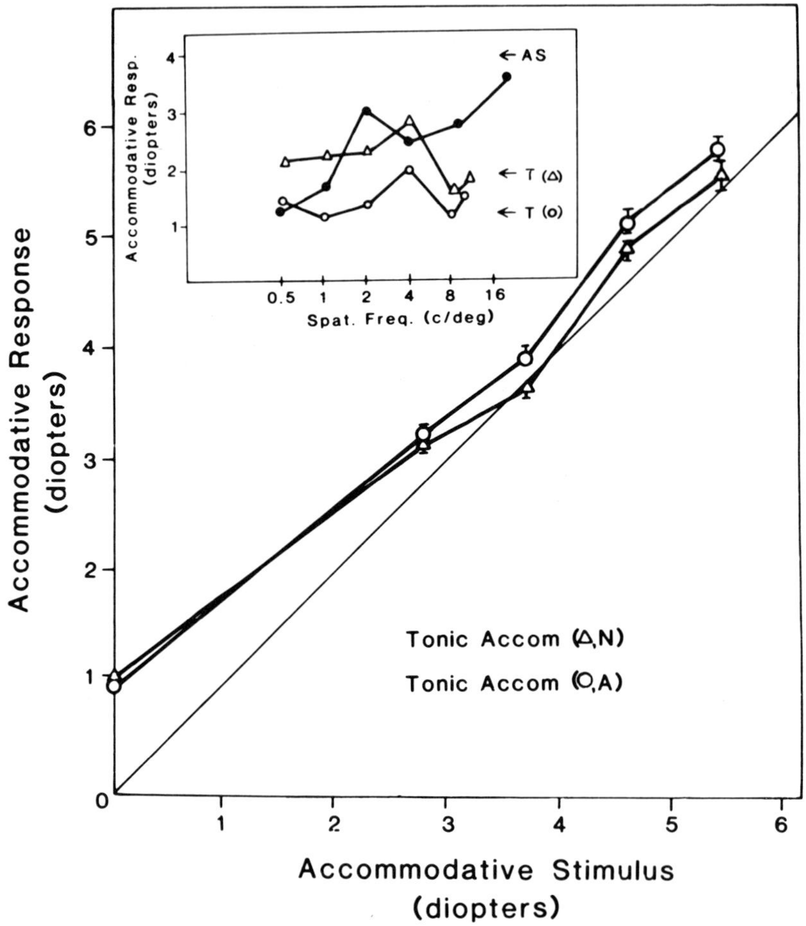

Fig. 15. Steady-state accommodation. Accommodative stimulus/response function near end of treatment period. Accommodation approximately equal and normal in the amblyopic (O) and dominant (**Δ**) eye. $\bar{X}$ 1 s.e.m. Inset shows accommodative spatial frequency response profile in the dominant eye (**Δ**), the amblyopic eye pre-treatment (O), and the amblyopic eye post-treatment (●). Accommodation measured using a Hartinger coincidence optometer; grating luminance 10 cd/m2; field of view 3.7 deg.

accommodative stimulus/response profile (Ciuffreda and Kenyon, 1983). Accommodation in each eye was within normal limits post-therapy; it was not assessed pre-therapy, but based on extensive testing of numerous other amblyopes, especially at this visual acuity level, markedly reduced accommodation would be predicted in the amblyopic eye (Ciuffreda et al, 1983). The second stimulus consisted of high contrast (80%), vertically-oriented sine-wave gratings presented at a 4 D stimulus level. Accommodation was measured as the grating spatial frequency was varied randomly. Accommodation was reduced over the entire range of spatial frequencies tested in the amblyopic eye as compared to the fellow dominant eye; however, following orthoptic therapy, accommodation in the amblyopic eye now exceeded that in the fellow dominant eye at all but the lowest spatial frequency. Further, accommodation to higher spatial frequencies was now possible with the amblyopic eye. Thus, steady-state accommodation to both sinusoidal and square-wave stimulus was markedly improved in the amblyopic eye following the lengthy and intensive orthoptic therapy.

DISCUSSION

Our results clearly demonstrate the considerable degree of visual system plasticity present in humans with functional amblyopia, even in those in whom it is thought by many that recovery of vision function is, for all practical purposes, impossible after the age of 6 years. This notion can probably be attributed to ophthalmologist Claude Worth (1903) who said, "After six years of age usually not much improvement in vision can be obtained, although I have met with many exceptions to the rule." Thus, even Worth himself allowed for some flexibility in this basic notion. At present, our general feeling, based on in-depth clinical and laboratory experience with about 30 older functional amblyopes, is that some degree of visual system plasticity is the rule rather than exception, at least up to the age of 30 years (Table 3). However, rate, degree, and pervasiveness of the recovery may indeed be somewhat diminished in the older patient. It should also be noted that the amblyopes who participated in our intensive conventional treatment program were not selected primarily on the basis of prognosis (which in most cases was poor). Instead, motivation, interest, dedication, and commitment to the rigorous treatment and test program, as well as ability to perform reliably in our psychophysical experiments, were our primary criteria. Further, we believe it is these factors, rather than etiology, age, or type of treatment plan, that were the common elements related to the successful therapeutic outcome. Lastly, improvement in the various vision functions that were tested always exceeded that which could be attributed to practice effects. For example, with respect to visual acuity (which is the most common measure used to monitor improvement with treatment), practice effects in a matched control group of amblyopes never exceeded $\pm$10% (Snell-Sterling visual efficiency) (Ciuffreda et al, 1980a). In contrast, treatment effects in our amblyopic patients always exceeded this limit, with improvements as great as 60% (e.g., 20/150 to 20/30) frequently occurring.

TABLE 3 --- Summary of Clinical and Laboratory Data on Orthoptic Effects in Older Amblyopes (Ages 6 to 30 yrs) For Various Vision Functions Tested By Us (top row, *) and By Others **

Visual Acuity	Accom. Stim./Resp. Function	Accom. Ampl.	Fixational Eye Move.	Contrast Sensitivity	Visual Evoked Response
8/11 (73%)	8/8 (100%)	5/5 (100%)	8/10 (80%)	4/4 (100%)	2/3 (66%)
28/35[a] (80%)			1/1[g] (100%)	2/26[h] (8%)	1/4[l] (25%)
19/23[b] (83%)				4/8[i] (50%)	1/1[m] (100%)
429/798[c] (54%)				9/9[j] (100%)	
7/7[d] (100%)				2/3[k] (66%)	
16/25[e] (64%)					
19/22[f] (86%)					

a = Selenow et al, in press
b = Cibis, 1978
c = Birnbaum et al, 1977
d = Kupfer, 1957
e = von Noorden et al, 1970
f = Brown and Edelman, 1976
g = von Noorden and Burian, 1958
h = Howell et al, 1983
i = Lennerstrand and Lundh, 1980
j = Sjostrand, 1981
k = Woo and Dalziel, 1981
l = Wilcox and Sokol, 1980
m = Barnard and Kelsey, 1980

* Not all patients were tested for all vision function listed.
** (number of patients showing improvement/number of patients tested)

Our research results are consistent with other important related findings including:

1. numerous clinical reports demonstrating marked improvement in visual acuity in the amblyopic eye of older patients following orthoptic therapy (Kasser and Feldman, 1953; Kupfer, 1957; Gould et al, 1970; von Noorden et al, 1970; Brown and Edelman, 1976; Flynn and Cassady 1978; Kivlin and Flynn, 1978; Selenow and Ciuffreda, 1983; see Birnbaum et al, 1977, for an excellent review), with apparent positive effects evident in patients up to 60 years of age.
2. several clinical reports demonstrating significant improvement in visual acuity in the amblyopic eye following either central vision loss or enucleation of the fellow dominant eye (Boase, 1951; Bishop, 1957; Gregersen, 1962; Kossowicz and Traczuk-Sorokowska, 1963; Tommila, 1964; Barnard and Kelsey, 1980; Rabin, 1984; Vereecken and Brabant, 1984), especially if intense (foveal) pleoptic treatment of the amblyopic eye is instituted; here again apparent positive effects were found in patients up to 60 years of age.
3. the paucity of laboratory investigations showing improved basic sensory function in amblyopic eyes following varying durations of orthoptic therapy in children and adult amblyopes (Flynn, 1968; Sjostrand, 1981; Thorn and Comerford, 1983), with effects evident in patients up to 40 years of age.
4. the paucity of experiments in adult cats showing partial post-critical period recovery and/or plasticity in the initially deprived eye following either enucleation of the dominant eye or reverse suturing (Creutzfeldt and Heggelund, 1975; Kratz et al, 1976; Hoffmann and Cynader, 1977; Singer et al, 1982).

It seems appropriate at this point to recall the general feelings of von Noorden and Crawford (1979) on this topic of plasticity. They stated that there is a sensitive or critical period wherein an impediment (such as strabismus or anisometropia) to the development of normal binocular vision producing abnormal visual experience (suppression and/or monocular contrast deprivation) early in life (up to the age of 6 years or so) frequently results in an anomaly of vision such as amblyopia. In contrast, there is a period of "physiological" reversal of such an abnormal visual experience-induced anomalous condition which extends much farther in years. Our own results show that considerable visual systems plasticity is possible in humans with functional amblyopia up to the age of 30 years, with the upper limit still undetermined.

Such plasticity in human, as is true for lower species such as cat and monkey, must reflect changes in the underlying neural substrate. Possible changes include release of neural inhibition or suppression from the dominant eye, reafferentation, growth of additional synapses, increase in synaptic gain, and/or increase in cell size or synaptic contact area. Additional work is essential to partition out those factors specifically involved in visual system plasticity in cat, monkey, and man. Such investigations will be facilitated by joint, multi-disciplinary efforts involving neurophysiologists, neuroanatomists, molecular biologists, and clinicians, with emphasis, whenever possible, on human material from patients with a well-documented history of amblyopia.

Finally, there is the need for a formal, large-scale investigation (perhaps a multi-center clinical trial as has been done for diabetic retinopathy) to study in detail in a clinically-controlled double-blind manner the effects of orthoptic therapy in human amblyopia using sophisticated psychophysical and electrophysiological assessment techniques, as well as careful clinic monitoring by expert clinicians, with emphasis in three areas. First there is the need to develop more efficient and effective treatment paradigms. At present, a "shotgun" approach is used involving a variety of procedures aimed at improving monocular and binocular vision function, with the optimal duration, and number of different procedures still poorly defined. In addition, there is the need to study the effects of age (of onset as well as at treatment), type of amblyopia, depth of amblyopia, degree of eccentric fixation, estimated age of onset of any strabismus, etc., on vision function recovery. While we may never reach a futuristic stage where a modular neural circuit board-type change can be made to replace the "amblyopic" circuit, neurochemical treatment (as has recently been advanced in the treatment of strabismus using botulism toxin; Scott, 1982) may become a reality in the foreseeable future (Duffy et al, 1978). At worst, the labor of such investigations may result in new ways to improve and maintain vision function in amblyopic eyes more easily, quickly, and inexpensively than is presently the case. The second area involves long-term follow-up of amblyopic patients of varying ages who have had initial significant recovery of vision. There is the need to determine those factors involved in the regression of vision found in some patients following termination of treatment. And third, it must be determined why some patients simply do not respond well to therapy, and tests must be developed that have high predictive value with respect to prognosis for recovery of vision in the amblyopic eye, to prevent wasted expenditure of time, effort, and money. While such goals may indeed seem lofty, they are not impossible if a concerted effort is made over the next decade or so by concerned clinicians and scientists.

ACKNOWLEDGMENTS

I would like to thank the following clinicians for performing much of the clinical testing and training, as well as providing intellectually stimulating discussions about the cases: Drs. M. Decker, S. Hokoda, D. Nathanson, E. Okimoto, D. Rumpf, C. Schnider, and A. Selenow. This work was supported in part by NIH Grant EY03541.

APPENDIX I --- Glossary of Clinical Terms

amblyopia - an anomaly of monocular vision in which there is reduced visual acuity which cannot be accounted for by uncorrected refractive error, structural abnormalities, and/or organic/neurological disease; it is generally attributed to the occurrence of early abnormal visual experience resulting from suppression in strabismus and contrast deprivation/suppression in anisometropia; it is present in about 2% of the general population.

anisometropia - a condition of unequal refractive state between the two eyes.

anomalous retinal correspondence - an anomaly of binocular vision in which the fovea of one eye and a non-foveal point in the other eye have identical visual direction; this is in contrast to normal retinal correspondence wherein the fovea of each eye has identical visual direction; it is present in most (up to 80%) strabismics.

eccentric fixation - an anomaly of monocular vision in which the time-average position of the fovea is off the object of regard, i.e., a non-foveal point is used for fixation; it is present in about 80% of amblyopic eyes.

orthoptics - a series of procedures and techniques aimed at attaining normal binocular vision; physiopsychological procedures for re-educating and reconditioning faulty levels of visual perception and for teaching the use of the two eyes together for efficient and confortable binocular vision. (Smith, 1950); it is frequently also called vision training.

strabismus - an anomaly of binocular vision in which the visual axis of one eye fails to intersect the object of regard; it is present in about 5% of the general population.

suppression - the process whereby all or part of the ocular image to one eye is prevented from contributing to the binocular percept.

REFERENCES

Allen, H. F. (1967) Incidence of amblyopia, Arch. Ophthalmol., 77:1.

Banks, M. S., Aslin, R. N. and Letson, R. D. (1975) Sensitive period for the development of human binocular vision, Science, 190: 675-677.

Barnard, W. M. and Kelsey, J. H. (1980) Changes in the visually evoked response during treatment for amblyopia in a uniocular patient, Trans. Ophthalmol. Soc. U.K., 100: 472-474.

Birnbaum, M. H., Koslowe, K., and Sanet, R, (1977) Success in amblyopia therapy as a function of age: a literature survey, Amer. J. Optom. Physiol. Optics, 54: 269-275.

Bishop, J. W. (1957) Treatment of amblyopia secondary to anisometropia, Brit. Orthopt. J., 14: 68-74.

Blakemore, C. and Van Sluyters, R. C. (1974) Reversal of the physiological effects of monocular deprivation in kittens: further evidence for a sensitive period, J. Physiol., 237: 195-261.

Blakemore, C., Vital-Durand, F. and Garey, L. J. (1981) Recovery from monocular deprivation in the monkey. I. Reversal of physiological effects in the visual cortex, Proc. Royal Soc. Lond. (B), 213: 399-423.

Blakemore, C., Hawken, M. J. and Mark, R. F. (1982) Brief monocular deprivation leaves subthreshold synaptic imput on neurones of cat's visual cortex, J. Physiol., 327: 489-505.

Boase, A. J. (1951) Amblyopia recovery at forty years of age, Brit. J. Ophthalmol., 35: 440-441.

Borish, I. M. (1970) Clinical Refraction, Professional Press, Chicago.

Brown, M. D. and Edelman, P. M. (1976) Conventional occlusion in the older amblyope, Amer. Orthopt. J., 26: 34-36.

Campbell, F. W., Hess, R. F., Watson, P. G. and Banks, R. (1978) Preliminary results of a physiologically based treatment of amblyopia, Brit. J. Ophthalmol., 62: 748-755.

Cibis, L. (1978) Amblyopia in microtropia, Metabol. Ophthalmol., 2: 143-145.

Ciuffreda, K. J., Kenyon, R. V. and Stark, L. (1979a) Saccadic intrusions in strabismus, Arch. Ophthalmol., 97: 1673-1679.

Ciuffreda, K. J., Kenyon, R. V. and Stark, L. (1979b) Different rates of functional recovery of eye movements during orthoptics treatment in an adult amblyope, Invest. Ophthalmol. Vis. Sci., 18: 213-219.

Ciuffreda, K. J., Kenyon, R. V. and Stark, L. (1979c) Abnormal saccadic substitution during small-amplitude pursuit tracking in amblyopic eyes, Invest. Ophthalmol. Vis. Sci., 18: 506-516..

Ciuffreda, K. J., Goldner, K. and Connelly R. (1980a) Lack of positive results of a physiologically based treatment of amblyopia, Brit. J. Ophthalmol., 64: 607-612.

Ciuffreda, K. J., Kenyon, R. V. and Stark, L. (1980b) Increased drift in amblyopic eyes, Brit. J. Ophthalmol., 64: 7-14.
Ciuffreda, K. J., and Hokoda, S. C. (1983) Spatial frequency dependence of accommodative responses in amblyopic eyes, Vis. Res., 23: 1585-1594.
Ciuffreda, K. J., Hokoda, S. C., Hung, G. K. and Semmlow, J. L. (1983) Accommodative stimulus/response function in human amblyopia, Doc. Ophthalmologica, 56: 303-326.
Ciuffreda, K. J. and Kenyon, R. V. (1983) Accommodative vergence and accommodatiion in normals, amblyopes, and strabismics, in Vergence Eye Movements: Basic and Clinical Aspects, Butterworth, Boston, pp. 101-173.
Ciuffreda, K. J. and Rumpf, D. (in press) Contrast and accommodation in human amblyopia, Vision Res.
Creutzfeldt, O. D. and Heggelund, P. (1975) Neural plasticity in visual cortex of adult cats after exposure to visual patterns, Science, 188: 1025-1027.
Crewther, D. P., Crewther, S. G. and Mitchell, D. E. (1981) The efficacy of brief periods of reverse occlusion in promoting recovery from the physiological effects of monocular deprivation in kittens, Invest. Ophthalmol. Vis. Sci., 21: 357-362.
Crewther, S. G., Crewther, D. P. and Mitchell, D. E. (1983) The effects of short-term occlusion on reversal of the anatomical and physiological effects of monocular deprivation in the lateral geniculate nucleus and visual cortex of kittens, Exp. Br. Res., 51: 206-216.
Duffy, F. H., Burchfiel, J. L., and Snodgrass, S. R. (1978) The pharmacology of amblyopia, Trans. Amer. Acad. Ophthalmol. Otolaryngol., 85: 489-495.
Duke-Elder, S. and Wybar, K. (1973) Ocular motility and strabismus, in System of Ophthalmology, vol. 6, Duke-Elder, S. (ed.), C. V. Mosby Co., St. Louis.
Flynn, J. T. (1968) Dark adaptation in amblyopia, Arch. Ophthalmol. 79: 697-704.
Flynn, J. T. and Cassady, J. C. (1978) Current trends in amblyopia therapy, Ophthalmology, 85: 428-450.
Flom, M. C., Weymouth, F. and Kahneman, D. (1963) Visual resolution and contour interaction, J. Opt. Soc. Amer., 53: 1026-1032.
Flom, M. C. (1966) New concepts on visual acuity, Optom. Wkly., 57: 63-68.
Flom, M. C. and Kerr, K. E. (1965) Amblyopia: a hidden threat?, J. Amer. Optom. Assoc., 36: 906-912.
Freeman, R. D. and Olson, C. R. (1979) Is there a "consolidation" effect for monocular deprivation? Nature, 282: 404-406.
Garey, L. J. and Vital-Durand, F. (1981) Recovery from monocular deprivation in the monkey. II. Reversal of morphological effects in the lateral geniculate nucleus, Proc. Royal. Soc. Lond. (B), 213: 425-433.
Garey, L. J. (1984) Structural development of the visual system of man. Human Neurobiol., 3: 75-80.
Gould, A., Fishkoff, D., and Galin, M. A. (1970) Active visual stimulation: a method of treatment of amblyopia in the older patient, Am. Orthopt., J., 20: 39-45.
Gregersen, E. (1962) Pleoptic treatment of remaining eye in father and son, Acta Ophthalmologica, 40: 385-388.
Griffin, J. R. (1976) Binocular Anomalies: Procedures for Vision Therapy, Chicago, Professional Press.

Harwerth, R. S. (1982) Behavioral studies of amblyopia in monkeys, Amer. J. Optom. Physiol. Optics, 59: 535-555.
Hillis, A., Flynn, J. T. and Hawkins, B. S. (1983) The evolving concept of amblyopia: a challenge to epidemiologists, Amer. J. Epidemiol., 118: 192-205.
Hoffmann, K. P. and Cynader, M. (1977) Functional aspects of plasticity in the visual system of adult cats after early monocular deprivation, Trans. Royal. Soc. Lond. (B), 278: 411-424.
Hoffmann, K. P. and Lippert, P. (1982) Recovery of vision with the deprived eye after the loss of the non-deprived eye in cats, Human Neurobiol., 1: 45-48.
Hokoda, S. C. and Ciuffreda, K. J. (in press) Different rates and amounts of vision function recovery during orthoptic therapy in an older strabismic amblyope, Ophthal. Physiol. Optics.
Howell, E. R., Mitchell, D. E., and Keith, C. G. (1983) Contrast thresholds for sine gratings of children with amblyopia, Invest. Ophthalmol. Vis. Sci., 24: 782-787.
Hubel, D. H. and Wiesel, T. N. (1965) Binocular interaction in striate cortex of kittens reared with artificial squint, J. Neurophysiol., 28: 1041-1059.
Jacobson, S. G., Mohindra, I. and Held, R. (1981) Development of visual acuity in infants with congenital cataracts, Brit. J. Ophthalmol., 65: 727-735.
Jones, K. R., Kalil, R. E. and Spear, P. D. (1984) Effects of strabismus on responsivity, spatial resolution, and contrast sensitivity of cat lateral geniculate neurons, J. Neurophysiol., 52: 538-555.
Kasser, M. D. and Feldman, J. B. (1953) Amblyopia in adults: treatment of those engaged in the various industries, Am. J. Ophthalmol., 36: 1443-1446.
Kivlin, J. D. and Flynn, J. T. (1981) Therapy of anisometropic amblyopia, J. Ped. Ophthalmol. Strab., 18: 47-56.
Kossowicz, H. and Traczuk-Sorokowska, N. (1963) Successful result of treatment of amblyopia in an adult, Klin. Oczna, 38: 411-414.
Kratz, K. E., Spear, P. D., and Smith, D. C. (1976) Post-critical period effects of monocular deprivation on striate cortex cells in the cat, J. Neurophysiol., 39: 501-511.
Kupfer, C. (1957) Treatment of amblyopia ex anopsia in adults, Am. J. Ophthalmol., 43: 918-922.
Lennerstrand, G. and Lundh, B. L. (1980) Improvement of contrast sensitivity from treatment for amblyopia, Acta Ophthalmol., 58: 292-294.
Levi, D. M. and Harwerth, R. S. (1982) Psychophysical mechanisms in humans with amblyopia, Amer. J. Optom. Physiol. Optics, 59: 936-951.
Levitt, F. B. and Van Sluyters, R. C. (1982) Recovery of binocular function in kitten visual cortex, J. Neurophysiol., 48: 1336-1346.
Malach, R. Ebert, R. and Van Sluyters, R. C. (1984) Recovery from effects of brief monocular deprivation in the kitten, J. Neurophysiol., 51: 538-551.
Mitchell, D. E. and Timney, B. (1984) Postnatal development of function in the mammalian visual system, in Handbook of Physiology, Section I, The Nervous System, vol. 3, Sensory Process, Part I, Darien-Smith, I. (ed.), Bethesda, Amer. Physiol. Assoc., pp. 507-555.
Movshon, J. A. and Van Sluyters, R. C. (1981) Visual neural development, Ann. Rev. Psychol., 32: 477-522.

Nathanson, D. R. and Ciuffreda, K. J. (1982) Results of intensive CAM grating treatment in a strabismic amblyope, Amer. J. Optom. Physiol. Optics, 59: 511-514.
Odom. J. V. (1983) Effects of visual deprivation on monocular acuities of humans and animals, Amer. J. Optom. Physiol. Optics, 60: 472-480.
Rabin, J. (1984) Visual improvement in amblyopia after visual loss in the dominant eye, Amer. J. Optom. Physiol. Optics, 61: 334-338.
Rauschecker, J. P. (1984) Neuronal mechanisms of developmental plasticity in the cat's visual system, Human Neurobiol., 3: 109-114.
Schapero, M. (1971) Amblyopia, Chilton Pub., Philadelphia.
Scott, A. B. (1982) Pharmacological treatment of strabismus, in Functional Basis of Ocular Motility Disorders, Lennerstrand, G., Zee, D. S. and Keller, E. L. (eds.), New York, Pergamon Press, pp. 199-205.
Selenow, A. and Ciuffreda, K. J. (1983) Vision function recovery during orthoptic therapy in an exotropic amblyope with high unilateral myopia, Amer. J. Optom. Physiol. Optics, 60: 659-666.
Selenow, A. and Ciuffreda, K. J. (in press) Vision function recovery during orthoptic therapy in an adult esotropic amblyope, Amer. J. Optom. Assoc.
Selenow, A., Ciuffreda, K. J., Mozlin, R., and Rumpf, D. (in press) Prognostic value of laser interferometric visual acuity in amblyopia therapy, Invest. Ophthalmol. Vis. Sci.
Sherman, S. M. and Spear, P. D. (1982) Organization of visual pathways in normal and visually deprived cats, Physiol. Rev., 62: 738-855.
Singer, W., Tretter, F. and Yinon, U. (1982) Evidence for long-term functional plasticity in the visual cortex of adult cats, J. Physiol., 324: 239-248.
Sjostrand, J. (1981) Contrast sensitivity in children with strabismic and anisometropic amblyopia: a study of the effect of treatment, Acta Ophthalmologica, 59: 25-34.
Smith, W. (1950) Clinical Orthoptic Procedures, C. V. Mosby, St. Louis.
Smith, D. C. (1981) Functional restoration of vision in the cat after long-term monocular deprivation, Science, 213: 1137-1139.
Swindale, N. V., Vital-Durand, F. and Blakemore, C. (1981) Recovery from monocular deprivation in the monkey. III. Reversal of anatomical effects in the visual cortex, Proc. Royal Soc. Lond. (B), 213: 435-450.
Thomas, J. (1978) Normal and abnormal contrast sensitivity functions in central and peripheral retinas, Invest. Ophthal. Vis. Sci., 17:746-753.
Thorn, F. and Comerford, J. P. (1983) Use of various measures of visual acuity and contrast sensitivity in the evaluation of monocular occlusion and active vision training in three amblyopes, Amer. J. Optom. Physiol. Optics, 60: 347-351.
Tieman, S. B. (1984) Effects of monocular deprivation on geniculocortical synapses in the cat, J. Comp. Neurol., 222: 166-176.
Tieman, S. B., Nickla, D. L., Gross, K., Hickey, T. L. and Tumosa, N.(1984) Effects of unequal alternating monocular exposure on the sizes of cells in the cat's lateral geniculate nucleus, J. Comp. Neurol., 225: 119-128.
Tommila, V. (1964) Treatment of amblyopia in loss of vision in the other eye, Acta Ophthalmologica, 42: 489-494.

Tommila, V. and Tarkkanen, A. (1981) Incidence of loss of vision in the healthy eye in amblyopia, Brit. J. Ophthalmol., 65: 575-577.

Vaegan and Taylor, D. (1979) Critical period for deprivation amblyopia in children, Trans. Ophthal. Soc. U. K., 99: 432-439.

Van Sluyters, R. C. (1978) Reversal of the physiological effects of brief periods of monocular deprivation in the kitten, J. Physiol., 284: 1-17.

Vereecken, E. P. and Brabant, P. (1984) Prognosis for vision in amblyopia after the loss of the good eye, Arch. Ophthalmol., 102: 220-224.

von Noorden, G. K., and Burian, H. M. (1958) An electro-oculographic study of the behavior of the fixation in amblyopic eyes in light and dark-adapted state: Am. J. Ophthalmol., 46: 68-77.

von Noorden, G. K., Springer, F., Romano, P., and Parks, M. (1970) Home therapy for amblyopia, Amer. Orthopt. J., 20: 46-50.

von Noorden, G. K. and Crawford, M. L. J. (1979) The sensitive period, Trans. Ophthal. Soc. U. K., 99: 442-446.

von Noorden, G. K., Crawford, M. L. J. and Levacy, R. A. (1983) The lateral geniculate nucleus in human anisometropic amblyopia, Invest. Ophthalmol. Vis. Sci., 24: 788-790.

Wiesel, T. N. and Hubel, D. H. (1963) Effects of visual deprivation on morphology and physiology of cells in the cat's lateral geniculate body, J. Neurophysiol., 26: 978-993.

Woo, G. C. and Dalziel, C. C. (1981) A pilot study of contrast sensitivity assessment of the CAM treatment of amblyopia, Acta Ophthalmol., 59:35-37.

Worth, C. A. (1903) Squint: Its Causes, Pathology and Treatment, P. Blakiston's Son and Co., Philadelphia.

Index